STUDY GUIDE FOR
MICROBIOLOGY
An Introduction

THIRD EDITION

Tortora • Funke • Case

Prepared by
Berdell R. Funke
North Dakota State University

The Benjamin/Cummings Publishing Company, Inc.
Redwood City, California • Fort Collins, Colorado
Menlo Park, California • Reading, Massachusetts • New York
Don Mills, Ontario • Wokingham, U.K. • Amsterdam • Bonn
Sydney • Singapore • Tokyo • Madrid • San Juan

ISBN 0-8053-0148-8

 BCDEFGHIJ-AL-8932109

The Benjamin/Cummings Publishing Company, Inc.
390 Bridge Parkway
Redwood City, California 94065

Table of Contents

Preface

To the Student

Welcome to microbiology. As with any subject, you will find some parts of it more interesting than others. But few subjects you will encounter in college touch on daily living in so many ways. You will find out things quite certain to interest you: how you become immune to diseases, how penicillin works, why oranges get moldy, and other "facts of life." On the other hand, now and then you may feel like the little girl thanking her aunt for a book on penguins. "Thank you very much," she wrote, "for the book on penguins. It told me a lot about penguins, even more than I wanted to know."

The Importance of Tutoring Yourself

I teach microbiology regularly to students required to take it for nursing, pharmacy, home economics, agriculture, civil engineering, and a number of other fields, including mortuary science. Not too long ago, one of the mortuary science students attempted the course and failed miserably. In none of three exams did he reach a score of 50%. At the end of the following term he told me, woefully, that he had just repeated the course and had missed a grade of C by two points. To transfer the credit to the school of mortuary science he needed at least a C.

It happens that our school has a policy of providing tutors for those who need them. I turned his case over to a bacteriology graduate student who tutored in his spacre time. On the first exam this formerly hapless undergraduate scored a 92 . . . and never looked back. Had three terms of exposure to my lectures finally enlightened him? Probably not; the tutor reported that in preparing for the first exam the student denied having ever heard of the procaryotic cell.

What, then, made the difference? The tutor did, of course, but remember that the tutor doesn't really put anything into your head that you cannot put there all by yourself. What the tutor does is force you to think about the material instead of just staring at it. The tutor exposes your areas of ignorance, but instead of letting it go at that, he forces you to pave over these areas with information.

You can do the same thing the tutor does—tutor yourself. The suggestions that follow and the use of this study guide will help you.

Lectures and Note-Taking

Instructors vary in their approach to microbiology courses. Some give lectures that largely paraphrase the text, and exams based mainly on the text. Others scarcely use the book at all in their lectures and base the exams on the lectures; successful students must be careful note-takers. (I have even heard of some cases in which there seems to be no apparent relationship between the information available to the student and the exams. These cases, of course, can only be left to the chaplain.)

For taking notes, it is important to put them down in an organized fashion such as this:

Staining of Bacteria
    Gram Stain
        Insert all material about Gram staining. Take down terms likely to be
        associated with Gram staining.
    Acid Fast Stain
        Insert all material about the acid fast stain. Take down all terms likely to be
        associated with acid fast staining.

Now, if you look back at notes like this, you can see that you were hearing about staining of bacteria and that there were basically two types of bacterial stains discussed. Note that you also clearly kept the information about the Gram stain separate from information about the acid fast stain. Never let your notes run together so that one topic merges with another. On objective tests this failure to separate discussions can be fatal. Make sure you have definitions straight. Never leave your notes or your mind a blank on something as obvious as the definition of a term.

Studying

I always invite students with grade problems in the course to come see me. I ask them to bring their notes and a copy of the offending exam, corrected from a key. Very often certain patterns show up. One is that the notes have no "study marks;" that is, no underlining or circling. Another is that the student often misses a cluster of questions pertaining to one group of topics, for example, the characteristics of different antibiotics. In other words, the student has encountered a list of antibiotics and failed to differentiate one from another. This is why it is important to organize information by general headings and a system of indentation as I have described. I also notice occasionally that a term, such as the word *autotroph*, stands alone in the notes without definition or explanation. Surely the student could have filled in such an omission by consulting the book or the instructor. In my presentation of the material, I am not trying to keep things secret, nor, presumably, is your instructor.

My recommendations are these: first, *organize your notes* so that it is as obvious as possible where a discussion of a topic, such as the disease diphtheria, has started, and make it clear where it ended. The most obvious way of doing this is to underline the main topic and inset all the information about that topic. (If you have trouble writing rapidly enough, or you have trouble with English as a language, ask the instructor if you can tape-record the lectures.)

The second point I make is that you should *never confuse staring at a page with studying a page*. As you read, do your eyes ever travel down half a page while your mind fails to take in a single word written on it? A person has only a limited capability for concentraion, and by staying up all night before an exam he or she is likely to exceed it. Do your studying in short bursts within the limits of your full attention span; start early. In order to ensure concentration, go through the notes or the text with a pencil or highlighting marker. Look deliberately at the material, and try to think like the person who goes through it to create an exam. You will find that your mind and the instructor's mind will often travel the same path. If the topic is penicillin, for example, mark the important facts: that it is produced by a mold, that its mode of action affects cell wall synthesis, that it has a beta-lactam ring in its structure, that it affects mainly gram-positive bacteria, and similar nuggets of information related to this antibiotic. When you have done this, you have actually *thought* about the material instead of just stared at it. This method will also teach you how to take notes. You will begin to recognize material that is likely to be on an exam and you will get it down; and you will simply leave to casual memory the instructor's anecdote about his memories of childhood chickenpox.

Study Guide Contents

This study guide provides a summary of each chapter that contains the important terms and concepts likely to be on examinations. This chapter summary is organized by the headings used in the text. Important terms are printed in boldface and defined, and important figures and tables from the text are included. Following each chapter synopsis is a sample exam on the chapter, an extensive self-testing section containing matching questions and fill-in questions. An answer key is provided. (You may wish to supplement these questions with the Study Questions at the end of each chapter in the textbook.) Obviously, not all questions can be anticipated, and a good instructor will usually introduce material other than that found in the text. But going through these sample exams will work as a self-tutoring device to direct attention to any areas of ignorance.

If you are going to improve your time in running the mile or the amount of weights you can lift, you must do more than read a book on techniques. You must apply them to build up your endurance or your muscles. The same applies to study habits. A guide cannot do your work for you; it can only serve as an tutor. You must do the work yourself.

1 *The Microbial World and You*

Microbes in Our Lives

Microbes include bacteria, fungi (yeasts and molds), protozoans, viruses, and the microscopic forms of algae. Most of these microorganisms are not harmful and indeed play a vital role in maintaining our global environment. Only a minority are pathogenic (disease producing). They are part of the food chain in oceans, lakes, and rivers; they break down wastes, incorporate nitrogen gas from the air into organic compounds, and participate in photosynthesis, which generates food and oxygen.

A Brief History of Microbiology

The First Observations

Anton van Leeuwenhoek was the first to report on the observation of microorganisms seen with magnifying lenses, beginning in 1674. He made detailed drawings of "animalcules" that have since been identified as representing bacteria and protozoans. About this time, **Robert Hooke** observed with a microscope the pores in slices of cork. He called them "cells." His discovery of the structure of cork, reported in 1665, marked the beginning of **cell theory**.

The Debate Over Spontaneous Generation

Until the second half of the nineteenth century, it was generally believed that life could arise spontaneously from nonliving matter, a process known as **spontaneous generation**. An early opponent of spontaneous generation, **Francesco Redi**, demonstrated in 1668 that maggots, the larvae of flies, do not arise spontaneously from decaying meat. He filled three jars with decaying meat and sealed them; three similar jars were left open. Maggots appeared only in the open vessels, which flies had entered. In later experiments, in which he covered the first three jars with gauze rather than sealing their tops, the results were the same. These later experiments demonstrated that access to air was not a factor.

Many, however, still believed that the simpler organisms observed by Leeuwenhoek might undergo spontaneous generation. In 1745, **John Needham** found that heated nutrient fluids poured into covered flasks were soon teeming with microorganisms. He took this as evidence of spontaneous generation. Twenty years later, **Lazzaro Spallanzani** showed that Needham's microorganisms had entered the fluid after boiling. Heating such fluids in a sealed flask, he showed, prevented the growth Needham had observed. Objections still remained; Needham felt that the heating had destroyed some vital force necessary for spontaneous generation. The concept of **biogenesis**, that living cells can arise only from other living cells, was introduced in 1858 by **Rudolf Virchow**.

In 1861, **Louis Pasteur** designed the experiments that finally ended the debate about spontaneous generation. He showed that flasks left open to the air after boiling would soon be contaminated, but if they were sealed, they remained free of microorganisms. He also used flasks whose long necks he bent into S-shaped curves. Air, with its presumed vital force, could enter these flasks, but airborne microorganisms were trapped in the tubes. The flask contents remained sterile. Pasteur showed that microorganisms are present throughout the environment and that they can be destroyed. He also devised methods to block access of airborne micro-organisms to nutrient environments; these methods were the basis of **aseptic** (germ-free) techniques, which are among the first things that a beginning microbiologist learns.

Fermentation and Pasteurization

At this time Pasteur was asked to investigate the problem of spoilage of beer and wine. Pasteur showed that, contrary to the belief that air acted on the sugars to convert them to alcohol, microorganisms called yeasts were responsible—and in the absence of air. This process is called **fermentation** and is used in making wine or beer. Spoilage occurs later when bacteria, in the presence of air, change the alcoholic beverage into acetic acid (vinegar). Pasteur prevented spoilage by heating the wine or beer just enough to kill such bacteria, a process that came to be known as **pasteurization**.

The Germ Theory of Disease

This association of yeasts with fermentation was the first concept to link a microorganism's activity to physical and chemical changes in organic materials. It suggested the possibility that microorganisms might be able to cause diseases as well—**the germ theory of disease**. In 1834, **Agostino Bassi** made the first association between a microorganism and a disease by proving that a silkworm disease was caused by a fungus. In 1865, Pasteur found that a silkworm disease was caused by a protozoan. Also in the 1860s, **Joseph Lister** applied the germ theory to medicine. He used carbolic acid (phenol) on surgical dressings and wounds, and greatly reduced the numbers of infections and deaths. In 1876, **Robert Koch** demonstrated that rod-shaped bacteria in the blood of cattle that had died of anthrax were the cause of death. He showed that these bacteria could be isolated and grown in pure culture, injected into healthy animals, and cause their death by anthrax. The same bacteria could then be isolated from dead animals. This demonstration, which proved that a specific microbe is the cause of a specific disease, followed a set of criteria known today as **Koch's Postulates**.

Vaccination

In 1798, **Edward Jenner** showed that the mild disease cowpox gave immunity to smallpox. He inoculated people with cowpox material by scratching their arm with a cowpox-infected needle. This process became known as **vaccination** ("vacca" being the Latin word for cow). The protection from disease provided by vaccination is called **immunity**. Years later, around 1880, Pasteur showed why vaccinations work. He found that the bacterium for chicken cholera lost its **virulence** (ability to cause disease) after it was grown for long periods in the laboratory. However, he showed that the weakened bacteria still retained their ability to induce immunity. Apparently the cowpox virus is related closely enough to smallpox to induce effective immunity.

The Birth of Modern Chemotherapy: Dreams of a "Magic Bullet"

The treatment of disease by chemical substances is called **chemotherapy**. When prepared from chemicals in the laboratory, these substances are called **synthetic drugs**, and when produced naturally by bacteria and fungi, they are called **antibiotics**. **Paul Ehrlich** speculated about a "magic bullet" that would destroy a pathogen without harming the infected host. In 1910, he found salvarsan, an arsenic derivative, that was effective against syphilis. Quinine, an extract of South American tree bark, had until then been the only other such chemical available. Spanish conquistadors used it to treat malaria. In the late 1930s, a survey of dye derivatives uncovered the important group of antibacterial sulfa drugs. The first antibiotic was discovered by **Alexander Fleming**, who observed the inhibition of bacterial growth by the mold *Penicillium notatum* and the

2

inhibitor penicillin. Penicillin was mass-produced and clinically tested in the 1940s. Since then, many antibiotics have been discovered.

Modern Developments in Microbiology

Immunology, the study of immunity, has expanded rapidly in the twentieth century. Smallpox has been eliminated and many new vaccines have become available. A major challenge will be to defeat the AIDS virus, which attacks the immune system.
Virology, the study of viruses, really began in 1892 when **Dmitri Iwanowski** demonstrated tobacco plant pathogens that would pass through filters too fine for known bacteria. Much later, **Wendell Stanley** showed that the organism, called the tobacco mosaic virus, was so simple and homogeneous it could be crystallized.

Naming and Classifying Microorganisms

The system of naming (**nomenclature**) we now use was established by **Carolus Linnaeus**. Scientific nomenclature assigns each organism two names. The **genus**, the first name, is always capitalized, and the **species** (specific epithet), which follows, is not capitalized. At first, organisms were grouped into either the animal kingdom or the plant kingdom. **H. R. Whittaker**, in 1969, devised a **five-kingdom classification** system, which groups organisms into **Procaryotae** (**Monera**), **Protista**, **Fungi**, **Plantae**, and **Animalia**, based on cellular organization and nutritional patterns.

The Diversity of Microorganisms

Bacteria

Bacteria are simple, one-celled organisms whose genetic material is not enclosed in a special nuclear membrane. For this reason, bacteria are called **procaryotes** (pre-nucleus); they make up the kingdom Whittaker calls Monera. Bacterial cells generally have one of three shapes: **bacillus** (rodlike), **coccus** (spherical or ovoid), and **spiral** (or corkscrew). Individual bacteria may form pairs, chains, or other groupings, which are usually the same within a species. Bacteria are enclosed in cell walls largely made of peptidoglycan (cellulose is the main substance of plant cell walls). Bacteria generally reproduce by binary fission into two equal daughter cells. Many move by appendages called flagella, and, although most use organic material for nutrition, some use inorganic substances or carry out photosynthesis.

Fungi

Fungi are **eucaryotes**—they contain DNA within a distinct nucleus, surrounded by a nuclear membrane. They may be unicellular or multicellular. Their cell walls are composed primarily of *chitin.* **Yeasts** are unicellular fungi larger than bacteria. **Molds** form *mycelia* of long filaments.

Protozoans

Protozoans are unicellular, eucaryotic microbes, members of the kingdom Protista. They are classified by their means of locomotion, such as *pseudopods, cilia,* or *flagella.*

3

Algae

Algae are photosynthetic eucaryotes, mostly of the kingdom Protista, and usually unicellular. They need light and air for growth.

Viruses

Viruses are very small and not cellular. They have a core of either DNA or RNA, which is surrounded by a protein coat. They may have a lipid envelope layer as well. They reproduce only inside the cells of the host organism.

Multicellular Animal Parasites

Flatworms and **roundworms** collectively called **helminths**, are not strictly micro-organisms. A part of their life cycles involve microscopic forms, however, and identifying them requires many techniques used in identifying traditional microorganisms.

Microbes and Human Welfare

Microbes recycle vital elements such as nitrogen, carbon, oxygen, sulfur, and phosphorus. *Cyanobacteria* (Monera formerly called blue-green algae) and certain soil bacteria may fix nitrogen from the air. In the carbon cycle, carbon dioxide is removed from the air by plants and algae, which convert it to food. Microorganisms decompose organic waste and return carbon dioxide to the atmosphere. Sewage treatment and pest control by insect pathogens are also important to human welfare.

Modern Industrial Microbiology and Genetic Engineering

In the future, several developments in industrial microbiology may become more important. **Single-cell protein**, a protein created by microorganisms such as yeasts, may be used directly as a food-substitute by humans. Also important will be **genetic engineering**, in which genes are transferred between different organisms in order to produce valuable products such as insulin, growth hormones, interferon, or vaccines. Hybrid DNA that contains genes from different organisms is called **recombinant DNA**.

Microbes and Human Disease

The relationship between microbes and disease will remain of great interest to us all. Our normal flora (microorganisms) do not disturb us and often are helpful. However, many diseases are caused by microorganisms. The study of the body's resistance to microbial infection and disease is a continuing part of microbiological research.

4

Test

In the matching section, there is only one answer to each question; however, the lettered options (a, b, c, etc.) may be used more than once or, perhaps, not at all.

I. Match

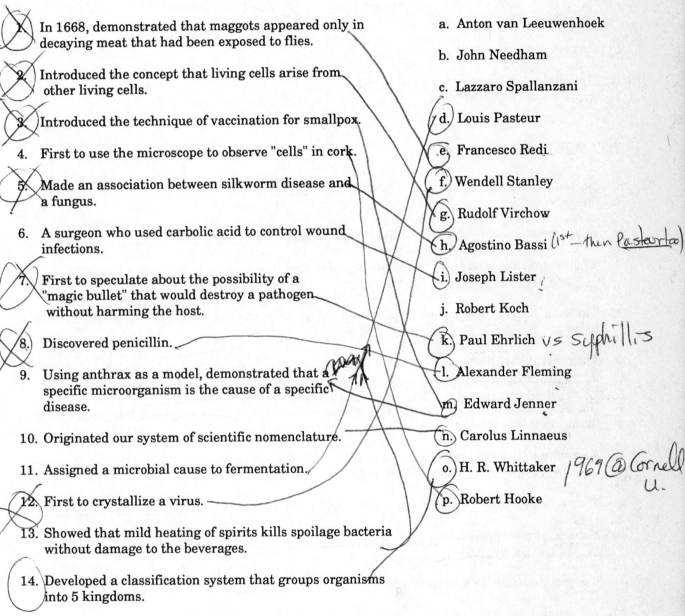

1. In 1668, demonstrated that maggots appeared only in decaying meat that had been exposed to flies.

2. Introduced the concept that living cells arise from other living cells.

3. Introduced the technique of vaccination for smallpox.

4. First to use the microscope to observe "cells" in cork.

5. Made an association between silkworm disease and a fungus.

6. A surgeon who used carbolic acid to control wound infections.

7. First to speculate about the possibility of a "magic bullet" that would destroy a pathogen without harming the host.

8. Discovered penicillin.

9. Using anthrax as a model, demonstrated that a specific microorganism is the cause of a specific disease.

10. Originated our system of scientific nomenclature.

11. Assigned a microbial cause to fermentation.

12. First to crystallize a virus.

13. Showed that mild heating of spirits kills spoilage bacteria without damage to the beverages.

14. Developed a classification system that groups organisms into 5 kingdoms.

a. Anton van Leeuwenhoek

b. John Needham

c. Lazzaro Spallanzani

d. Louis Pasteur

e. Francesco Redi

f. Wendell Stanley

g. Rudolf Virchow

h. Agostino Bassi (1st — then Pasteur too)

i. Joseph Lister

j. Robert Koch

k. Paul Ehrlich vs Syphillis

l. Alexander Fleming

m. Edward Jenner

n. Carolus Linnaeus

o. H. R. Whittaker 1969 @ Cornell U.

p. Robert Hooke

II. Match

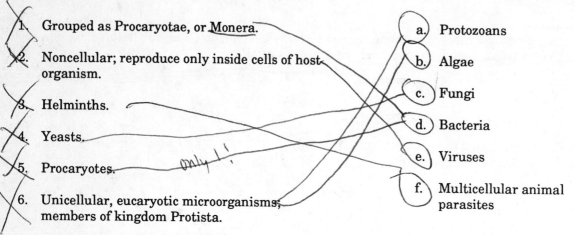

1. Grouped as Procaryotae, or <u>Monera</u>.
2. Noncellular; reproduce only inside cells of host organism.
3. Helminths.
4. Yeasts.
5. Procaryotes.
6. Unicellular, eucaryotic microorganisms; members of kingdom Protista.

only 1!

a. Protozoans
b. Algae
c. Fungi
d. Bacteria
e. Viruses
f. Multicellular animal parasites

III. Fill-in-the-Blanks

1. Bacteria generally reproduce by a process called _____binar fiss_____ into two equal daughter cells.

2. Microorganisms classified as photosynthetic eucaryotes are _____Algae_____.

3. Techniques that keep areas free of unwanted microorganisms are referred to as _____aseptic_____ (germ-free) techniques.

4. The set of criteria that prove a specific microorganism is the cause of a specific disease is known today as _____Koch's postulates_____

5. The concept that living cells can arise only from other living cells is called _____biogenesis_____.

6. One objection proponents of spontaneous generation made to experiments with heating nutrient fluids in sealed containers was that heating destroyed some __"life component"__ in the air. *vital*

7. According to the rules applied to the scientific naming of a biological organism, the _____genus_____ is always capitalized.

8. The protection from a disease that is provided by vaccination is termed ___immunity___.

9. Monera, formerly called blue-green algae, which may fix nitrogen from the air, are called _____Cyanobacteria_____

10. The general name for a rod-shaped bacterium is _____bacillus_____.

11. The general name for a spherical or ovoid bacterium is _____coccus_____.

12. Protists that are classified by their means of locomotion are _____protozoa_____.

13. The process by which yeasts change sugars into alcohol is termed _____fermentation_____.

14. The treatment of a disease with chemical substances is called _chemotherapy_

15. Paul Ehrlich discovered an arsenic derivative, _Salvarsan_, that was effective against syphilis.

16. Antimicrobial chemicals produced naturally by bacteria and fungi are called _antibiotics_

Key

Matching

I. 1.e 2.g 3.m 4.p 5.h 6.i 7.k 8.l 9.j 10.n 11.d 12.f 13.d 14.o
II. 1.d 2.e 3.f 4.c 5.d 6.a

Fill-in-the-Blanks

1. binary fission   2. algae   3. aseptic   4. Koch's Postulates   5. biogenesis
6. vital force   7. genus   8. immunity   9. *Cyanobacteria* 10. bacillus   11. coccus
12. protozoans   13. fermentation   14. chemotherapy   15. salvarsan   16. antibiotics

2 *Chemical Principles*

Structure of an Atom

All matter is made up of small units called **atoms**. Atoms in certain combinations form **molecules**. All atoms have a centrally located **nucleus** and particles called **electrons**, which have a negative (-) charge, moving around the nucleus. The nucleus of an atom is made up of positively (+) charged particles called **protons** and uncharged particles called **neutrons**. Both neutrons and protons have about the same weight, 1840 times that of an electron. Since the total positive charge on the nucleus equals the total negative charge of the electrons, each atom is an electrically neutral unit. Different kinds of atoms are listed by **atomic number**, the number of protons in the nucleus. The **atomic weight** is the total number of protons and neutrons in an atom.

Chemical Elements

All atoms with the same atomic number are classified as the same **chemical element**. Examples of elements are hydrogen (H), carbon (C), sodium (Na), nitrogen (N), and sulfur (S). About 26 of the 92 naturally occurring elements are found commonly in living things. Most elements have **isotopes**, forms of the element, which have the same number of protons in the nucleus but different weights because of differing numbers of neutrons. For example, oxygen isotopes have the same atomic number, 8, but different atomic weights $^{16}_{8}O$ $^{17}_{8}O$ $^{18}_{8}O$.

Electronic Configurations

Electrons are arranged in **electron shells**, which are regions corresponding to different **energy levels** (Table 2-1). The arrangement is called an **electronic configuration**. The chemical properties of atoms are largely a function of the number of electrons in the outermost electron shell. When the outer shell of an atom is filled, as in helium, it is stable, inert; it does not react with other atoms. Partially filled electron shells make for atoms that tend to react with other atoms to become more stable.

How Atoms Form Molecules: Chemical Bonds

The **valence**, or combining capacity of an atom, is dependent on the number of extra electrons or deficient electrons in its outermost electron shell. Hydrogen has a valence of one and carbon has a valence of four. This means that hydrogen can form only one chemical bond and carbon four. When atoms gain stability by completing the full complement of electrons, they form **molecules**. A molecule containing at least two kinds of atoms, such as water (H_2O), is a **compound**. The attractive forces holding molecules together are chemical bonds. In general, atoms form bonds by: 1) gaining or losing electrons from the outer electron shell, or 2) sharing outer electrons.

Ionic and Covalent Bonds

Bonds formed by the first method are **ionic bonds**; those formed by the second are **covalent bonds**.

Sodium chloride (NaCl) is an example of a substance formed by ionic bonding. Sodium (Na) has a single electron in the outer shell, which it tends to lose; it is an *electron donor*. The atom is positively charged. This atom reacts strongly with chlorine, which is deficient one electron of outer shell capacity. Chlorine has a negative charge because it usually fills this vacancy with electrons from other atoms; that is, it is an *electron acceptor*. **Cations** such as potassium (K^+) or sodium (Na^+), are positively charged ions; their outer electron shell is less than half filled and they lose

Table 2-1 Electronic Configurations for the Atoms of Some Elements Found in Living Systems

Element	Electronic Configuration			Diagram	Number of Valence (Outermost) Shell Electrons
	First Electron Shell	Second Electron Shell	Third Electron Shell		
Hydrogen	1				1
Carbon	2	4			4
Nitrogen	2	5			5
Oxygen	2	6			6
Magnesium	2	8	2		2
Phosphorus	2	8	5		5
Sulfur	2	8	6		6

electrons. **Anions** are negatively charged ions; with their outer shells *more* than half filled, they tend to gain electrons. Examples of anions are iodine (I^-), chlorine (Cl^-), and sulfur (S^{2-}) ions. Covalent bonds are stronger and more common in organisms than ionic bonds.

Hydrogen Bonds

The **hydrogen bond** consists of a hydrogen atom covalently bonded to one oxygen or nitrogen atom (these being the most commonly involved elements). In water, for example, the electrons are all closer to the oxygen nucleus than to the hydrogen nucleus. As a result, the oxygen portion of the molecule has a slight negative charge and the hydrogen portion, a slight positive charge. The hydrogen bond is a result of the water molecule's slightly positive polar charge being attracted to the negative end of the other molecules. Although these bonds are relatively weak, large molecules may have hundreds of them.

Molecular Weight and Moles

The **molecular weight** of a molecule is the sum of the atomic weights of all its atoms. A **mole** is the number of grams equal to the molecular weight.

Chemical Reactions

The making or breaking of bonds between atoms are **chemical reactions**. The process of two or more atoms, ions, or molecules combining to form new and larger molecules is called a **synthesis reaction**. The combining substances are **reactants** and the **product** is the substance formed. All synthesis reactions are collectively called **anabolism**; an example is the formation of starch from sugar molecules. The reverse of synthesis is a **decomposition reaction**; an example is the digestion of food. Collectively, these reactions are called **catabolism**. **Exchange reactions** are partly synthesis and partly decomposition. In theory, all reactions are **reversible reactions**, but in practice some reverse more easily than others.

Energy of Chemical Reactions

Synthesis *requires* energy (an **endergonic reaction**) and decomposition *yields* energy (an **exergonic reaction**). Energy used by organisms for synthesis is provided by food digestion or photosynthesis.

How Chemical Reactions Occur

The minimum collision energy required for a chemical reaction to occur, according to **collision theory**, is its **activation energy**. This is the amount of energy needed to disrupt the stable electron configuration of a molecule so a new arrangement can occur. Increases in temperature, pressure, and concentration increase collision frequency and increase **reaction rates**.

Enzymes and chemical reactions. **Enzymes** are large protein molecules that have a three-dimensional **active site** tailored to interact with a specific **substrate**. They act as **catalysts**, to accelerate chemical reactions. The enzyme-substrate complex that is formed lowers the activation energy required, and thus enables collisions to be more effective. Enzymes speed up reactions without an increase in temperature—important in biological systems.

Important Biological Molecules

Water

Inorganic compounds such as water are essential for living cells. **Water**, a **polar molecule**, has an unequal charge distribution that gives it four important characteristics: 1) a high boiling point and, incidentally, a solid form (ice) that is less dense than the liquid, a characteristic that allows it to float; 2) excellent solvent properties; 3) an important role as a reactant or product in many chemical reactions; and 4) an excellent temperature buffer capability that helps protect the cell from changes in environmental temperature.

Acids, Bases, and Salts

When inorganic salts such as NaCl are dissolved in water, they dissociate, or ionize, into ions. Acids dissociate into one or more hydrogen ions (or protons H^+) and one or more negative ions. In other words, acids are a proton (H^+) donor; HCl, for example, yields H^+ and Cl^-. A base, such as NaOH, dissociates into Na^+ and and the **hydroxyl ion** (OH^-). A salt is a substance that dissociates into ions that are neither H^+ nor OH^-. NaCl is an example; it dissociates into Na^+ and Cl^-.

Acid-Base Balance

The **pH** of a solution is the negative logarithm to the base 10 of the hydrogen ion concentration in moles per liter. Acidic solutions contain more H^+ ions than OH^- ions and have a pH lower than 7. Alkaline, or basic, solutions have more OH^- ions than H^+ ions. A pH of 7, with equal concentrations of H^+ and OH^-, is neutral. Each change of number on the logarithm pH scale represents a tenfold change in concentration. **Buffers** are compounds that keep the pH from changing drastically.

Organic Compounds

All **organic compounds** contain carbon whose four covalent bonds allow it to combine with neighboring or other atoms to form large structures. Hydrogen fills most of the free bonds in these structures. Many other **functional groups** such as **hydroxy** (OH) and **amino** (NH_2) groups are also found. Such fundamental groups can be added to **carbon skeletons** to make alcohols or amino acids. Small organic molecules can be combined into very large molecules called **macromolecules**. Generally, these are **polymers**, which are formed by repeating **subunits** of **monomers**. When two monomers join together, they usually produce a molecule of water, a reaction called **condensation** or **dehydration synthesis**.

Carbohydrates

Carbohydrates have the general formula of CH_2O, with a hydrogen-oxygen ratio of 2 to 1. Carbohydrates form **monosaccharides**, simple sugars of three to seven carbon atoms; **disaccharides**, two monosaccharides bonded together; and **polysaccharides**, eight or more monosaccharides such as glucose joined by dehydration.

Lipids

Lipids also are composed of atoms of carbon, hydrogen, and oxygen, but they do not have the 2 to 1 hydrogen-oxygen ratio. They are diverse in structure but share the property of being soluble in nonpolar solvents such as ether and alcohol, but not water. Examples are **fats**, which are formed from glycerol and fatty acids joined by an **ester link**. **Saturated fats** have no double bonds; **unsaturated fats** have several. Important in the membranes of cells are **phospholipids**, which also contain phosphorus, nitrogen, and sulfur. Lipids called **sterols** (members of the steroids) are important constituents in the plasma membranes of cells.

Proteins

Amino acids are the subunits of **protein**, which are organic molecules containing carbon, hydrogen, oxygen, and nitrogen, as well as some sulfur. Amino acids have at least one carboxyl (-COOH) group and one amino (-NH$_2$) group attached to the same carbon atom, the alpha carbon—hence the name **alpha-amino acid**. Amino acids exist in configurations called **stereoisomers**, which are designated D or L. (Only L-isomers are found in biological proteins, except for those found in some bacterial cell walls and in some antibiotics.) Only 20 different kinds of amino acids occur naturally in proteins. Amino acids in proteins are connected by **peptide bonds**.

Proteins have four levels of organization: **primary**, the order in which amino acids are linked (their sequence); **secondary**, one-dimensional coiling or zigzag arrangements which form helixes and pleated sheets, and are held together by hydrogen bonding; **tertiary**, three-dimensional shapes in which sulfhydryl groups can form covalent disulfide bonds, and hydrogen bonds provide the holding force; and a **quaternary** structure, which is an aggregation of two or more individual polypeptide units. An example of the last is an enzyme called phosphorylase, which consists of four identical polypeptide units, none of which exhibits enzymatic activity by itself. **Conjugated proteins** are formed by combinations of amino acids and other organic or inorganic components. **Glycoproteins**, for example, contain sugars, **nucleoproteins** contain nucleic acids, and **lipoproteins** contain lipids.

Nucleic Acids

The basic units of **nucleic acids** are **nucleotides**. There are two principal nucleic acids; **deoxyribonucleic acid (DNA)** and **ribonucleic acid (RNA)**. Each nucleotide unit of DNA contains three parts; a nitrogen-containing base, a five-carbon sugar (**deoxyribose**) and a phosphoric acid molecule (Figure 2-1). The nitrogen-containing base is either **adenine, guanine, cytosine**, or **thymine**. Adenine and guanine are double-ring structures, **purines**. Thymine and cytosine are single-ring structures, **pyrimidines**. Note in the figure how adenine always pairs with thymine and cytosine with guanine. This **complementary pairing** allows one strand to be reproduced from the structure of the other. RNA differs from DNA in several ways. It is usually single-stranded, the five-carbon sugar is **ribose**, and **uracil** replaces thymine. **Nucleosides** have a purine or pyrimidine attached to a pentose sugar but have no phosphate group.

Adenosine Triphosphate (ATP)

The principal energy-carrying molecule in all cells is **adenosine triphosphate (ATP)**. It consists of an adenosine unit of adenine and ribose joined to three phosphate groups (Figure 2-2). It releases much energy when converted to **adenosine diphosphate (ADP)** by the addition of a water molecule, which separates the terminal phosphate group. ATP is generated by energy-yielding reactions in the cell, such as the decomposition of glucose. This energy carried in ATP can be used to perform synthesis.

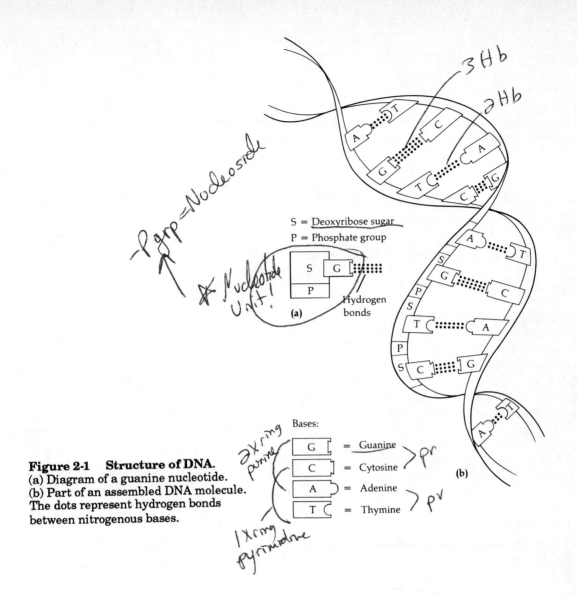

[handwritten annotations: "3 Hb", "2 Hb", "P grp = Nucleoside", "# Nucleotide Unit!"]

S = Deoxyribose sugar
P = Phosphate group

Hydrogen bonds

(a)

[handwritten: "2X ring purine", "1X ring pyrimidine"]

Bases:

G	= Guanine
C	= Cytosine
A	= Adenine
T	= Thymine

[handwritten: "> Pr", "> Pv"]

(b)

Figure 2-1 Structure of DNA.
(a) Diagram of a guanine nucleotide.
(b) Part of an assembled DNA molecule.
The dots represent hydrogen bonds
between nitrogenous bases.

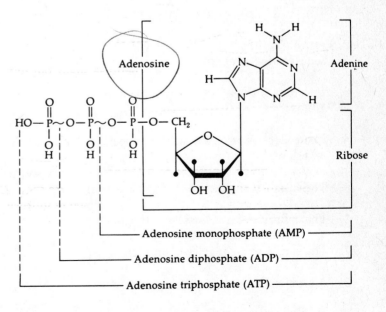

Figure 2-2 Structure of ATP. High-
energy phosphate bonds are wavy. When
ATP breaks down to ADP and inorganic
phosphate, a large amount of chemical
energy is released that can be used in other
chemical reactions.

Adenosine

Adenine

Ribose

Adenosine monophosphate (AMP)

Adenosine diphosphate (ADP)

Adenosine triphosphate (ATP)

Test

In the matching section, there is only one answer to each question; however, the lettered options (a, b, c, etc.) may be used more than once or, perhaps, not at all.

I. Match

1. The strongest of the three chemical bonds listed. *h*   a. Proton

2. An uncharged particle in the atomic nucleus. *f*   b. Atomic weight

3. A hydrogen ion. *a*   c. Electron

4. The number of protons in the nucleus. *e*   d. Nucleus

5. Particles with a negative charge, which move in shells around the nucleus. *c*   e. Atomic number

6. A bond formed by sharing electrons in the outermost shell. *h*   f. Neutron

and know? 7. A weak bond formed, for example, by the slight positive charge at the hydrogen end of the water molecule reacting with the negative end of other molecules. *i*   g. Ionic bond

   h. Covalent bond

   i. Hydrogen bond

8. A bond formed by the gain or loss of electrons from the outer electron shell. *g*

II. Match

1. The substance upon which an enzyme acts. *c*   a. Cation *pos*
   b. Anion *negtv*
2. A protein that lowers the activation energy required for a reaction. *j*   c. Substrate
   d. Valence
   e. Mole
3. The sum of the atomic weights of a molecule's atoms. *f*   f. Molecular weight
   g. Anabolism
   h. Catabolism
4. The collective term for all decomposition reactions. *h*   i. Mole
   j. Enzyme
5. The number of grams equal to molecular weight. *e* ok! same answr

6. The collective term for all synthesis reactions. *g*

7. The combining capacity of an atom. *d*

8. An ion with a positive charge. *a*

14

III. Match

1. Prevents drastic change in pH. *g*

2. Substance that dissociates into ions that are neither (OH⁻) or (H⁺). *~~a~~ d*

   ?2nd know

3. A proton donor. *~~a~~ c*

4. Dissociates into one or more negative ions such as (OH⁻). *e*

5. Combinations of atoms that have gained stability by completing the full complement of electrons in the outermost shell. *~~a~~ b*

a. Compound
b. Molecule
c. Acid
d. Salt
e. Base
f. Hydroxyl ion
g. Buffer
h. Hydroxy group

IV. Match

1. Eight or more glucose molecules in a chain. *d*

2. Sterol. *e*

3. Fat. *e*

4. Production of a molecule of water during synthesis *a*

5. Formed from chains of amino acids. *f*

6. Lipoprotein. *g*

7. Results from release of energy by separation of terminal phosphate group. *j*

8. DNA *h*

a. Condensation
b. Monosaccharides
c. Disaccharides
d. Polysaccharides
e. Lipids
f. Proteins
g. Conjugated proteins
h. Nucleic acids
i. Adenosine triphosphate
j. Adenosine diphosphate
k. Sterol

V. Fill-in-the-Blanks

1. All atoms with the same atomic number are classified as the same _chem elem_.

2. When discussing synthesis, the combining substances are called _react's_ and the substance formed is the _prod_.

3. Carbon has a valence of _4_.

4. Most elements have _isotopes_, which have the same number of protons in the nucleus but different weights due to differing numbers of neutrons.

5. Amino acids in proteins are connected by _peptide b's_.

15

6. The five-carbon sugar in DNA is __deoxyribose__.

7. In DNA, adenine always pairs with __thymine__.

8. In RNA, thymine is replaced by __uracil__.

9. The principal energy-carrying molecule in all cells is __ATP__.

10. RNA differs from DNA in being usually __1X__ stranded.

11. In a protein the order of the amino acid sequence is the __1^o__ level of organization.

12. The basic units of nucleic acids are __nucleotides__.

13. Thymine and cytosine are __1X__ ring structures called __pyrimidines__.

14. An example of a nitrogen-containing base in a nucleotide is __ATCG/U__. (more than one answer is acceptable)

15. The __3^o__ level of protein organization provides it with a three-dimensional shape.

16. About __20__ different kinds of amino acids occur naturally in proteins.

17. Amino acids can exist as __stereo isomers__, designated D or L.

18. Some important characteristics of water are its high __polarity, boil pt__, and its capacity as a temperature __buffer__. (?for hydroly s/dehydrat-condensat) ie wh brk dwn cmplex sug ar → 1X

2nd know

19. A molecule with at least two kinds of atoms, such as water, is a __compound__.

20. Cations are positively charged ions; their outer electron shell is __less__ than half filled and they lose electrons.

21. The minimum collision energy required for a chemical reaction to occur is its __activat E__.

22. __Carbo's__ have the general formula of CH_2O, with a hydrogen-oxygen ratio of 2 to 1.

23. Sucrose, formed of the sugars glucose and fructose, is an example of a __di__ saccharide. (provide the prefix)

24. __Lipids__ share the property of being soluble in nonpolar solvents such as ether and alcohol, but not water.

?nd kn 25. Neutrons and protons have a weight about 1840 times that of __e⁻s__.

?nd knw 26. Decomposition yields energy, which is called an __e.__ reaction.

Key

Matching

I. 1.h 2.f 3.a 4.e 5.c 6.h 7.i 8.g
II. 1.c 2.j 3.f 4.h 5.e 6.g 7.d 8.a
III. 1.g 2.d 3.c 4.e 5.b
IV. 1.d 2.e 3.e 4.a 5.f 6.g 7.j 8.h

Fill-in-the-Blanks

1. chemical element   2. reactants; product   3. four   4. isotopes   5. peptide bonds
6. deoxyribose   7. thymine   8. uracil   9. adenosine triphosphate (ATP)   10. single-
11. primary   12. nucleotides   13. single-; pyrimidines
14. adenine, guanine, cytosine, thymine, uracil   15. tertiary   16. 20
17. stereoisomers   18. boiling point; buffer   19. compound   20. less
21. activation energy   22. carbohydrates   23. di-   24. Lipids   25. electrons
26. endergonic — ? should be exergonic — Release And E

17

3 *Observing Microorganisms Through a Microscope*

Units of Measurement

Microorganisms are measured by metric units unfamiliar to many of us. The **micrometer** (μm), which formerly was known as the **micron**, is equal to 0.000001 meter. The prefix *micro* indicates that the unit following should be divided by one million. A **nanometer** (**nm**), formerly known as a **millimicron** (mμ) is equal to 0.000000001 meter. *Nano* tells us that the unit should be divided by one billion. An **angstrom** (Å) is equal to 0.0000000001 meter.

Microscopy: The Instruments

Compound Light Microscopy

The **compound light microscope** has two sets of lenses; the **objective** and the **ocular**. Specimens magnified by the objective lens—magnified 100 times, for example—are remagnified by the ocular, usually 10 times. Thus, the total magnification is 1000 times. Most microscopes provide magnifications of 100, 400, and 1000. A magnification of 2000 times is about the highest obtainable. The specimen is illuminated by visible light that is passed through a **condenser**, which directs the light rays through the specimen (Figure 3-1a). **Resolution**, or **resolving power**, is the ability of a microscope to distinguish between two points. The shorter the wavelength of the illumination, the better the resolution.

For the highest magnification, it is necessary to use **oil immersion objectives**. Immersion oil has the same **refractive index** as glass; that is, the relative velocity of light passing through it is the same. Without immersion oil filling the space between the slide bearing the specimen and the objective, the image will be fuzzy with poor resolution (Figure 3-1b).

Darkfield Microscopy

Some microorganisms, such as the thin spirochete, *Treponema pallidum*, which causes syphilis, are best seen with **darkfield microscopy**. In the darkfield microscope, an opaque disk blocks light from entering the objective directly. The light hits only the sides of the specimen, and scattered light enters the objective and reaches the eyes. The specimen appears white against a black background.

Phase-Contrast Microscopy

Living microorganisms do not show up well in the ordinary compound light microscope because they have a refractive index similar to water. The **phase-contrast microscope** takes advantage of subtle differences in the refractive index of different parts of the living cell and its surrounding medium. As light is slowed down in the denser portions, these differences cause certain structures to stand out. The microorganism and many of its internal structures are seen in the natural state, alive and unstained.

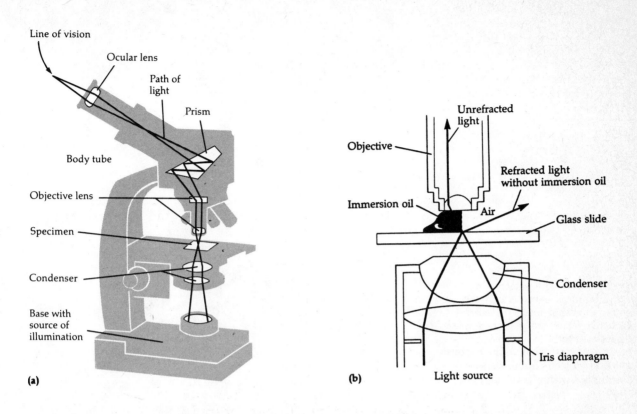

Figure 3-1 The compound light microscope. **(a)** The path of light. **(b)** Refractive index. Because the refractive indexes of the glass microscope slide and immersion oil are the same, the light rays do not refract when passing from one to another.

Fluorescence Microscopy

Certain fluorochrome dyes, which glow with visible light—yellow, for example—when illuminated by ultraviolet light, can be used to view and identify microorganisms. **Fluorescence microscopy** techniques use a special microscope with ultraviolet light illumination; this light illuminates the specimen but is not permitted to reach the eye. The stained microorganism—for example, *Mycobacterium tuberculosis* (the bacterium causing tuberculosis) or *Bacillus anthracis* (the agent causing anthrax)—is highly visible against a dark background in such a microscope. However, the principal use of these dyes and microscopes is in the **fluorescent-antibody** technique, or **immunofluorescence**. In this technique, the organisms are allowed to react on a slide with antibodies (highly specific proteins produced by the body's defense system). A fluorescent dye is attached to the **antibody**. The combination of the antibody, the attached dye, and the microorganism for which the antibody is specific (called an **antigen**; it stimulates the body to produce these antibodies) allows the microorganism's presence to be detected (Figure 3-2). Because the antibody is specific for a particular microorganism, this is a very useful diagnostic technique, and is often used for syphilis and rabies.

Electron Microscopy

The wavelengths of electrons, which travel in waves much as light does, are only about 1/100,000 as long as those of visible light, and therefore have much better resolving power.

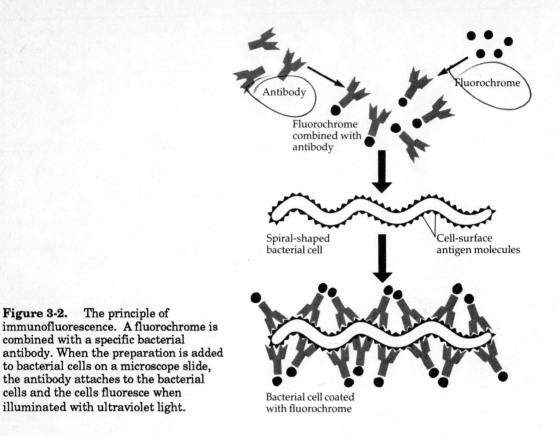

Figure 3-2. The principle of immunofluorescence. A fluorochrome is combined with a specific bacterial antibody. When the preparation is added to bacterial cells on a microscope slide, the antibody attaches to the bacterial cells and the cells fluoresce when illuminated with ultraviolet light.

Labels within figure: Antibody; Fluorochrome; Fluorochrome combined with antibody; Spiral-shaped bacterial cell; Cell-surface antigen molecules; Bacterial cell coated with fluorochrome

Transmission Electron Microscopy. In **transmission electron microscopy**, a beam of electrons is passed through ultrathin sections of the specimen and focused on a fluorescent screen, where it is visible to the eye and can be photographed. Magnifications of about 200,000 times are possible. Contrast can be enhanced by staining with the salts of metals such as lead, tungsten, and uranium.

Scanning Electron Microscope. Three-dimensional views of specimens at magnifications of about 10,000 times are possible with **scanning electron microscopy**. Here, the electron beam is directed at the intact specimen from the top, rather than passing through a section, and electrons leaving the surface of the specimen are viewed on a televisionlike screen. Spectacular pictures of intact organisms are possible.

Preparation of Specimens for Light Microscopy

Staining and Smear Preparation

Most microorganisms are viewed in stained preparations; that is, they are colored with a dye to make them visible or to emphasize certain structures. A thin film of a microbial suspension, called a **smear**, is spread on the surface of a slide. Flaming the air-dried smear coagulates the microbial proteins and fixes the microorganisms to the slide so they do not wash off. The smear can now be stained. **Basic dyes** have a colored ion that is positive, which helps them adhere to bacteria, which are slightly negative. Examples of basic dyes are crystal violet, methylene blue, and safranin. **Acidic dyes**, having a negative color ion, are more attracted to the back-ground than to the negatively charged bacteria; thus, a field of colorless bacteria is presented against a stained background. This is an example of **negative staining**.

20

Simple Stains

To visualize shapes and arrangement of cells, a **simple stain** is usually sufficient. A chemical to make the microorganisms stain more intensely, a **mordant**, may be added.

Differential Stains

The most useful **differential stain** is the **Gram stain**, developed by Hans Christian Gram. It divides bacteria into two large groups: **gram-positive** and **gram-negative**. The first step in preparing a Gram stain is to apply a purple dye, crystal violet, to a heat-fixed smear. After a water rinse, an iodine mordant is then added. When a smear stained in this manner is washed with ethanol or an ethanol acetone solution, some species of bacteria are decolorized and others are not. If the smear retains the purple dye, the organism is gram-positive. If the alcohol removes the dye, the colorless microorganisms are no longer visible. Safranin, a red dye, is then applied and the decolorized, or gram-negative bacteria, appear pink. Safranin is used here as a counterstain. The Gram stain reflects a basic difference in the cell wall structure of bacteria. It is a first step in identification, and the susceptibility of microorganisms to antibiotics is often related to the Gram reaction.

Acid-Fast Stain. Members of the genera *Mycobacterium* (the cause of tuberculosis and leprosy) and *Nocardia* possess a cell wall and cell membrane that appear to have an attraction for the red dye carbolfuchsin. Therefore, the **acid-fast stain**, in which carbolfuchsin is applied and gently steamed for several minutes, will stain them red. This dye is held so firmly that the cells are not decolorized by acid-alcohol, which does remove the dye from bacteria that are not acid-fast. A methylene blue counterstain will make a slide in which acid-fast organisms are red and others are blue. The acid-fast stain is an invaluable aid in the diagnosis of tuberculosis.

Special Stains

A colloidal suspension of dark particles such as India ink can be used as a **capsule stain**. The capsule will appear around each bacterial cell as a halo from which the India ink carbon particles are excluded. Endospores do not stain by ordinary methods, but the **Schaeffer-Fulton endospore stain**, which uses malachite green as a primary stain and safranin as a counterstain, shows endospores as green within red or pink cells. Flagella are too small to be resolved by light microscopes. In a **flagella stain**, a mordant can be used to increase the diameter of the flagella until they are visible in a light microscope.

Test

In the matching section, there is only one answer to each question; however, the lettered options (a, b, c, etc.) may be used more than once or, perhaps, not at all.

I. Match

1. The electrons pass through a thin section of the *d* specimen.

2. Visible light passes through specimen; uses *a* separate objective and ocular lenses.

3. Details become visible because of differences in *c* the refractive index of different parts of the cell.

4. Visible light is scattered from the sides of the *f* specimen and the specimen is visible against a darkened background.

5. A special microscope using ultraviolet illumination. *e*

6. The electrons strike the surface of the specimen, *b* and those leaving the surface are viewed on a television-like screen.

a. Compound light microscope

b. Scanning electron microscope

c. Phase-contrast microscope

d. Transmission electron microscope

e. Fluorescence microscope

f. Darkfield microscope

II. Match

1. Pertaining to the relative velocities of light *e* through a substance.

2. Involves the use of antibodies and ultraviolet light. *g*

3. One millionth of a meter. *a*

4. One ten-millionth of a meter. *c*

5. The ability to separate two points in a microscope field. *d*

a. Micrometer
b. Nanometer
c. Angstrom
d. Resolving power
e. Refractive index
f. Immersion oil
g. Immunofluorescence

22

III. Match

1. Adhere(s) best to bacteria, which have a negative charge, because the color molecule has a positive charge. *a*

2. Used in diagnosis of tuberculosis. ⊗ *d mycobacter*

3. Involve(s) use of a negative stain made from India ink particles. ⊗ *e*

   malach green +boil

4. Schaeffer-Fulton stain. ⊗ *f*

5. Use(s) carbolfuchsin dye. *d*

6. Use(s) malachite green. *f*

7. Reflect(s) a basic difference between microbial cell walls; ethanol will not remove stain from some bacteria. *c*

a. Basic dyes

b. Acidic dyes

c. Gram stain

d. Acid-fast stain

e. Capsule stain

f. Spore stain

g. Flagella stain

IV. Fill-in-the-Blanks

1. *Nano* says that the unit should be divided by a ___billion___.

2. About the highest magnification possible in a compound light microscope is ___1000 (2000)⊗___

3. Immersion oil has about the same refractive index as ___glass___.

4. Fluorochrome dyes glow with visible light when illuminated by ___UV lt___.

5. Electron wavelengths are only about ___1/100,000___ as long as visible light and therefore have much ___greater___ resolving power.

6. Bacteria tend to have a slightly ___negtv___ electrical charge.

7. The electron microscope that tends to give a three-dimensional view of the specimen is the ___scan___ electron microscope.

8. The thin film of a microbial suspension spread on the surface of a slide is called a ___smear___.

9. Flaming the slide before applying the stain is called ___ht fx___.

10. Transmission electron microscopy permits magnifications as high as about ___200,000 X___.

11. In the flagella stain, a ___mordant___ is used to increase the diameter of the flagella.

23

12. Bacterial genera that are acid-fast are _____ *Mycobacterium* _____ and _____ *Nocardia* _____ .

13. A disease for which the acid-fast stain is useful in diagnosis is _____ TB / Leprosy _____ .

14. In order to see shapes and arrangements of cells, a _____ simple _____ stain is usually sufficient.

15. A chemical to make microorganisms stain more intensely is called a _____ Mordant _____ .

16. _____ A _____ dyes have a negative color ion.

17. Some microorganisms such as the thin spirochete, *Treponema pallidum*, are best seen with _____ darkfld _____ microscopy.

Key

Matching

I. 1.d 2.a 3.c 4.f 5.e 6.b
II. 1.e 2.g 3.a 4.c 5.d
III. 1.a 2.d 3.e 4.f 5.d 6.f 7.c

Fill-in-the-Blanks

1. billion   2. 2000   3. glass   4. ultraviolet light   5. 1/100,000; better
6. negative   7. scanning   8. smear   9. fixing   10. 200,000 times   11. mordant
12. *Mycobacterium*; *Nocardia*   13. tuberculosis or leprosy   14. simple
15. mordant   16. Acidic   17. darkfield

4 *Functional Anatomy of Procaryotic and Eucaryotic Cells*

Procaryotic and Eucaryotic Cells

Distinguishing characteristics of **procaryotic** (*prenucleus*) cells are:

1. Genetic material not enclosed within a membrane;
2. Lack other membrane-bounded organelles;
3. No histone proteins associated with DNA;
4. Cell walls usually contain peptidoglycan; and
5. Usually divide by binary fission.

Distinguishing characteristics of **eucaryotic** (*true nucleus*) cells are:

1. Nucleus is bounded by a membrane;
2. DNA of chromosomes usually associated with proteins called histones and non-histones; and
3. Possess mitotic aparatus, mitochondria, endoplasmic reticulum, and sometimes chloroplasts.

The Procaryotic Cell

Size, Shape, and Arrangement of Bacterial Cells

Bacteria range in size from 0.2 to 2.0 μm in diameter. Basic bacterial shapes are: the spherical **coccus** (meaning "berry"), the rod-shaped **bacillus** (meaning "little staff"), and the spiral. **Diplococci** form pairs; **streptococci** form chains; tetrads divide in two planes, forming groups of four; **sarcinae** divide in three regular planes and form cubelike packets; **staphylococci** divide in irregular, random planes and form grapelike clusters. Most bacilli are single rods, but they can appear in pairs—**diplobacilli**—or in chains—**streptobacilli**. Coccobacilli are ovals. **Vibrios**, slightly curved, commalike rods, are also included among spiral bacteria. **Spirilla** have a helical corkscrew shape, and are motile by means of flagella. **Spirochetes** are shaped like spirilla but have an axial filament for motility. **Pleomorphic** bacteria have an irregular morphology.

Structures External to the Cell Wall

Glycocalyx

The general term for substances surrounding bacterial cells is **glycocalyx, (extracellular polymeric substance)**, which is usually a polysaccharide or polypeptide. If organized and tightly attached, it is called a **capsule** (Figure 4-1). The glycocalyx aids in attachment to surfaces; capsules contribute to pathogenicity by protecting from phagocytosis.

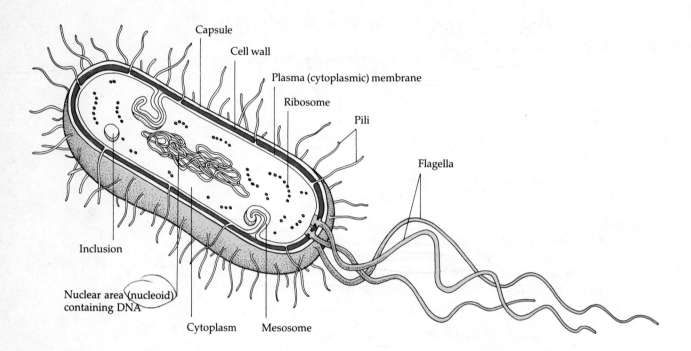

Capsule
Cell wall
Plasma (cytoplasmic) membrane
Ribosome
Pili
Flagella
Inclusion
Nuclear area (nucleoid) containing DNA
Cytoplasm   Mesosome

Figure 4-1   Structure of a typical procaryotic (bacterial) cell as seen in longitudinal section.

Flagella

Flagellar filaments are composed of a protein, **flagellin**. The base of the flagellar filament widens to a **hook**. Attached to the hook is a **basal body** (a rod with rings), which anchors the flagellum to the cell wall and plasma membrane. (See Figure 4.2.) The basal body of gram-negative bacteria is anchored to the cell wall and plasma membrane; in gram-positive bacteria, it is anchored only at the plasma membrane. Flagella, when present, are arranged in one of four basic ways: **monotrichous** (single polar flagellum); **lophotrichous** (two or more polar flagella at one end of cell); **amphitrichous** (tufts of flagella at both ends of the cell); **peritrichous** (flagella distributed over the cell). Flagella may spin clockwise or counter-clockwise, producing directional movement or random tumbling.

Axial Filaments

Spirochetes move by means of **axial filaments**, which arise near cell poles inside the cell wall, and wrap spiral fashion around the cell. These can cause the spirochetes to move in a snakelike manner.

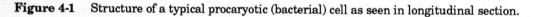

Pili

Many bacterial cells have numerous hairlike appendages called **pili** or **fimbriae** that are shorter than flagella. They help the cell adhere to surfaces such as mucous membranes—often a factor in pathogenicity. Pili are shorter and thinner than flagella. Specialized pili, **sex pili**, assist in the transfer of genetic material between bacteria.

26

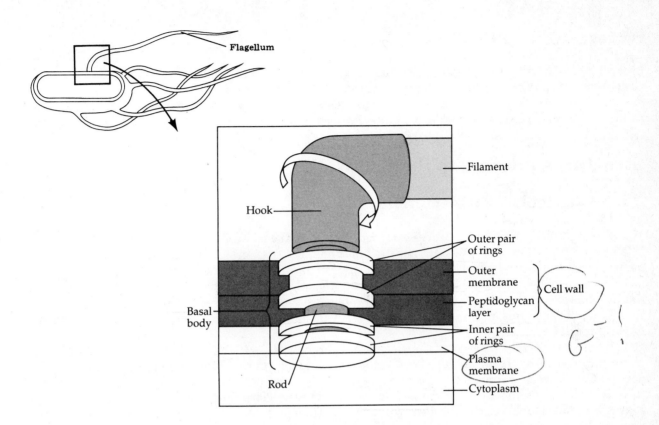

Figure 4-2 Flagella. Parts and attachment of a flagellum of a gram-negative bacterium.

Cell Wall

The bacterial cell wall is a semirigid structure giving the characteristic shape of the cell.

Composition and Characteristics

The cell wall of gram-positive bacteria is composed of **peptidoglycan** (murein), which consists of two sugars—N-acetylglucosamine and N-acetylmuramic acid—and also chains of amino acids. The two sugars alternate with each other, forming a carbohydrate backbone. Peptide side chains of four amino acids attached to the N-acetylmuramic acid are cross-linked to form the macromolecule of the cell wall. Many gram-positive bacteria also contain polysaccharides called **teichoic acids**. The cell wall of gram-negative bacteria also contains peptidoglycans, but in smaller amounts. These cells have a lipoprotein, lipopolysaccharide (**LPS**), and a phospholipid **outer membrane** surrounding their peptidoglycan layer. A periplasmic space is found between the outer membrane and the plasma membrane. The outer membrane also provides resistance to phagocytosis, and when the cell disintegrates in the host's bloodstream, the lipid portion of the LPS (**Lipid A**) is released as an **endotoxin** that can cause illness. Materials may penetrate the outer membrane through **porins**, (small and nonspecific), and **specific channel proteins** (very specific and which serve as attachment points for viruses and bacteriocins).

   Mycoplasma bacteria do not have cell walls. They are unique also in having sterols in their plasma membranes. **L forms** also are bacteria without cell walls, but they are mutants and may eventually revert to the original form with a cell wall.

Damage to the Cell Wall

Lysozyme, an enzyme occurring in tears, mucus, and saliva, damages cell walls of many gram-positive bacteria. A bacterium that has lost its cell wall and is surrounded only by the plasma membrane is a **protoplast**. Gram-negative cells treated with lysozyme retain much of the outer membrane layer and are called **spheroplasts**. Both are sensitive to osmotic shock.

Structures Internal to the Cell Wall

Plasma (Cytoplasmic) Membrane

The **plasma (cytoplasmic) membrane** is just internal to the cell wall and encloses the cytoplasm. It is about 60% protein and 40% lipid (mostly phospholipid). Both procaryotic and eucaryotic membranes have a two-layered structure, molecules in parallel rows, called a **phospholipid bilayer**. One end (phosphate) is water soluble and the other (hydrocarbon) is insoluble. The water soluble ends are on the outside of the bilayer. Protein molecules are embedded in the membrane and they, and phospholipids, may move freely within the membrane; this arrangement is called the **fluid mosaic model**.

The most important function of the plasma membrane is as a selective barrier. It is **selectively permeable (semipermeable)** and certain molecules and ions pass through and others do not. Several factors affect permeability. Large molecules such as proteins cannot pass; smaller molecules such as amino acids and simple sugars can pass if uncharged. (The phosphate end of the bilayer is charged.) Lipid soluble substances, because of the phospholipid content, pass more easily. Plasma membranes contain enzymes that help break down nutrients and produce more energy. The **chromatophores** or **thylakoids**, which contain pigments for bacterial photosynthesis, are found in the plasma membranes.

Mesosomes are folds in the plasma membrane that are involved in binary fission. First a cross wall, called a **transverse septum**, forms and partitions the genetic material. Mesosomes begin the formation of the transverse septum and may help separate DNA into each daughter cell following binary fission. Mesosomes also increase the area of the plasma membrane, aiding nutritional efficiency.

Movement of Materials Across Membranes

Material crosses plasma membranes by passive processes such as **simple diffusion** (movement of molecules or ions from an area of higher concentration to an area of lower concentration). At equilibrium, the concentration gradient has been eliminated. **Osmosis** is the force with which a solvent (such as water) moves from a solution of lower solute concentration (such as dissolved sugar) to a solution of higher solute concentration. **Isotonic** solutions have equal solute concentrations on both sides of the membrane. **Hypotonic** solutions have a lower concentration of solutes outside the cell than inside; this is the case with most bacteria. **Hypertonic** solutions have a higher concentration of solutes outside the cell, and bacterial cells placed in such solutions lose water by osmosis and shrink, and the cytoplasm collapses within the cell wall. A third passive process is **facilitated diffusion**, in which a carrier protein (**permease**) combines with and transports a substance across the membrane, but only where a concentration gradient is present. **Active transport** requires cell energy (ATP) and also involves carrier proteins moving substances across the plasma membrane. In **group translocation**, the substance is chemically altered during transport. Once inside, the plasma membrane is impermeable. This is important for low concentration substances.

Cytoplasm and Inclusions

The term **cytoplasm** means everything inside the plasma membrane. It has many inclusions such as **metachromatic granules** of stored phosphate (**volutin**), **polysaccharide granules** of glycogen and starch, **lipid inclusions** such as poly-ß-hydroxybutyric acid, and sulfur granules. The cytoplasm also contains many **ribosomes**. **Carboxysomes** are inclusions found in bacteria that use carbon dioxide as their sole source of carbon. **Gas vacuoles** or vesicles help some bacteria maintain buoyancy.

Nuclear Area. The **bacterial chromosome**, which contains the genetic information, is a single, long, circular molecule of DNA found in the **nuclear area**, or **nucleoid**. Small cyclic DNA molecules, **plasmids**, are not connected to the chromosome. **Plasmids** do not contain normally essential genes but may provide a selective advantage under abnormal conditions—antibiotic resistance, for example.

Endospores

Endospores are highly resistant bodies formed by a few bacterial species, such as *Bacillus* and *Clostridium*. **Sporulation** or **sporogenesis** is the process of their formation. First, there is an ingrowth of the plasma membrane (**spore septum**). A small portion of the cytoplasm and newly replicated bacterial chromosome is then surrounded by a membrane, the **forespore**. A thick **spore coat** of protein forms around this membrane. The endospore core is dehydrated and contains considerable **dipicolinic acid**, as well as a few essential materials necessary to return it to its vegetative state, which is accomplished through the process of **germination**.

The Eucaryotic Cell

Flagella and Cilia

Eucaryotic **flagella** are relatively long; **cilia** are more numerous and are shorter. Both are involved in locomotion, and both contain small tubules of protein called **microtubules**.

Cell Wall

Most algae and some fungi have walls containing cellulose, and often fungi have chitin as well. Yeast cell walls contain the polysaccharides glucan and mannan. No eucaryotic cell wall contains peptidoglycans.

Plasma (Cytoplasmic) Membrane

In eucaryotic cells, the **plasma membrane**, which contains sterols, may be the external cell covering. Substances cross the membrane by mechanisms similar to procaryotes. In addition, a sort of engulfment, **endocytosis** brings particles, even some viruses, into the cell. Examples are **phagocytosis**, used by white blood cells to ingest (destroy) bacteria (Chapter 15) and **pinocytosis** (cell drinking) by which liquids enter cells.

Cytoplasm

Cytoplasm is the matrix in which various cellular components are found. The complex internal structure of **microfilaments**, **intermediate filaments**, and **microtubules** is called the **cytoskeleton**, all held together by the **microtrabecular lattice**.

Organelles

In eucaryotes, unlike procaryotes, many important enzymes are found in and functions carried out by **organelles**.

Nucleus

The **nucleus** is an oval organelle containing the DNA. It is surrounded by a **nuclear envelope**. **Pores** in the nuclear membrane allow the nucleus to communicate with the **endoplasmic reticulum** of the cytoplasm. The **nucleoplasm** is a gellike fluid in the nucleus. **Nucleoli**, which may be the center for the synthesis of ribosomal RNA, are present.

Endoplasmic Reticulum and Ribosomes

The **endoplasmic reticulum** (**ER**) is a network of canals running throughout the cytoplasm. It may provide a surface area for chemical reactions, a pathway for transporting molecules, and a storage area for synthesized molecules; it has a role in lipid and protein synthesis. **Ribosomes** are sites of protein synthesis in the cell.

Golgi Complex

The **Golgi complex** consists of four to eight flattened channels connected to the ER. Its function is largely secretion (removal from the cell) of proteins, lipids, and carbohydrates. The Golgi complex also functions in carbohydrate synthesis, including glycoproteins.

Mitochondria

Mitochondria are organelles which have a smooth outer membrane and an inner membrane that is arranged in a series of folds called **cristae**. The center of the mitochondrion is called the **matrix**. Enzymes forming ATP are located on the cristae—which is one reason why mitochondria are called powerhouses of the cell.

Chloroplasts

Photosynthesizing cells contain membrane-bounded structures called **chloroplasts**, which contain chlorophyll and enzymes involved in photosynthesis. The chlorophyll is found in stacks of membranes called **thylakoids**. Stacks of thylakoids are **grana**.

Lysosomes

White blood cells, which destroy bacteria by phagocytosis, contain many **lysosomes**. These are membrane-enclosed spheres containing powerful digestive enzymes. When injured, cells may enzymatically destroy themselves, hence the name "suicide packets" for lysosomes.

Centrioles

Centrioles are bundles of microtubules that probably play a role in cell division.

Table 4-1 Principal Differences Between Procaryotic and Eucaryotic Cells

Characteristic	Procaryotic	Eucaryotic
Size of cell	Typically 1 - 10 μm in diameter *sm*	Typically 10-100 μm in diameter
Nucleus	No nuclear membrane or nucleoli	True nucleus consisting of nuclear membrane and nucleoli
Membrane-bounded organelles	Absent	Present; examples include lysosomes, Golgi complex, endopolasmic reticulum, mitochondria, and chloroplasts
Flagella	Consist of two protein building blocks	Complex; consist of multiple microtubles
Glycocalyx	Extracellular polymeric capsule or slime layer	Absent
Cell wall	Usually present; chemically complex	When present, chemically simple *plants*
Plasma membrane	No carbohydrates and generally lacks sterols	Sterols and carbohydrates that serve as receptors are present
Cytoplasm	No cytoskeleton or streaming *except Mycoplasm*	Cytoskeleton; cytoplasmic streaming
Ribosomes	Small size (70S)	Large (80S); small (70S) in organelles
Chromosome (DNA) arrangement	Single circular chromosome; lacks histones	Several or many linear chromosomes with histones
Cell division	Binary fission	Mitosis
Sexual reproduction	No meiosis; exchange of DNA fragments only	Involves meiosis

Evolution of Eucaryotes

According to the **autogenous hypothesis** the specialized internal membranes evolved into the various organelles. The **endosymbiotic hypothesis** proposes that organelles such as mitochondria and chloroplasts, especially, arose from evolution of ingested procaryotic cells. Table 4-1 outlines the principal differences between procaryotic and eucaryotic cells.

Test

In the matching section, there is only one answer to each question; however, the lettered options (a, b, c, etc.) may be used more than once or, perhaps, not at all.

I. Match

1. Helical; move by flagella, if present. ~~h~~

2. Spherical; in chains. *c*

3. Divide in three regular planes; spheres form cubelike packets. *a*

4. Helical; axial filament for motility. *d*

5. A simple, commalike curve. *e*

6. Name means "little staff." *f*

7. Ovals. *j*

a. Sarcinae
b. Tetrads
c. Streptococci
d. Spirochetes
e. Vibrios
f. Bacilli
g. Cocci
h. Spirilla
i. Diplococci
j. Coccobacilli

II. Match

1. Golgi complex. *a*

2. Meiosis occurs in reproduction. *a*

3. Single circular chromosome without histones. *b*

4. Sterols generally present in cell membrane. *a*

5. Cell wall almost always contains peptidoglycans. *b*

6. Nucleus bounded by a membrane. *a*

a. Eucaryotic cell
b. Procaryotic cell

III. Match

1. Contain pigments for photosynthesis by bacteria; found in the plasma membrane.

2. Gram-negative bacterial cell walls after their treatment with lysozyme. *g*

3. Specialized pili that assist in the transfer of genetic material between cells. *d*

4. Numerous hairlike appendages that help attachment to mucous membranes. *c*

5. Polysaccharide capsules formed by many bacteria to help them adhere to environmental surfaces. *a*

6. Polysaccharides found in the cell wall of many gram-positive bacteria. *f*

a. Glycocalyx
b. Flagellin
c. Pili
d. Sex pili
e. Capsules
f. Teichoic acids
g. Spheroplasts
h. Protoplasts
i. Chromatophores
j. Chloroplasts

IV. Match

1. Metachromatic granules of stored phosphate in procaryotes. *a*

2. Enzyme storage banks in eucaryotic cells. *d*

3. Entrance of fluids into eucaryotic cells. *j*

4. Membrane-enclosed spheres in phagocytic cells that contain powerful digestive enzymes. *g*

5. The "powerhouses" of the cell. *h*

6. A gellike fluid found in the eucaryotic nucleus. *f*

7. An inner membrane found in the mitochondria. *c*

a. Volutin
b. Plasmids
c. Cristae
d. Zymogens
e. Ribosomes
f. Nucleoplasm
g. Lysosomes
h. Mitochondria
i. Phagocytosis
j. Pinocytosis

33

V. Match

1. Arrangement of flagella distributed over the entire cell. _f_

2. Arrangement of tufts of flagella at both ends of cell. ✗ _g_

3. A widening at the base of the flagellar filament. _e_

4. An enzyme affecting gram-positive cell walls; found in tears, etc. _d_

5. A compound found in bacterial endospores. _b_

6. A compound found frequently in the cell walls of yeasts. _c_

a. Exocytosis
b. Dipicolinic acid
c. Chitin
d. Lysozyme
e. Hook
f. Peritrichous
g. Amphitrichous
h. Lophotrichous   ✗ @ 1 end
i. Monotrichous
j. Flagellin

VI. Match

1. Closely involved in protein synthesis. _d_

2. Structure(s) characteristic of both eucaryotic and procaryotic plasma membranes. _a_

3. Found in the flagella and cilia of eucaryotic cells. _c_

a. Phospholipid bilayer
b. Transverse septum
c. Microtubules
d. Ribosomes

VII. Fill-in-the-Blanks

1. Chemically the capsule is a ___poly sacchar___, a polypeptide, or both.

2. Capsules protect pathogenic bacteria from ___phagocy___, a process by which protective host cells engulf and destroy microorganisms.

3. ___Plasmids___ are small cyclic DNA molecules that are not connected with the main chromosome.

4. In the chloroplasts of eucaryotic cells, the chlorophyll is found in stacks of membranes called ___. Thylakoids___

5. ___ in the nuclear membrane allow the nucleus to communicate with the endoplasmic reticulum of the cytoplasm. _pores_

6. ___(endo)spores___ are highly resistant bodies formed by a few bacterial species.

7. In binary fission, first, a cross wall, the transverse ___septum___, forms and partitions the genetic material.

8. A bacterium that normally forms a cell wall but has lost this ability is called an ___protoplast___. _L form_

9. In the sequence of spore formation, first there is an ingrowth of the plasma membrane called the ___mesosome___ *[spore septum]*. A small portion of the cytoplasm is then surrounded by a membrane forming the ___forespore___.

10. Ribosomes are attached to the endoplasmic reticulum in areas that are granular, called the _____. *[rough reticulum]*

11. The center portion of the mitochondrion is called the_____. *[matrix]*

12. Bundles of microtubules, which probably play a role in cell division and are located within centrosomes, are called _____. *[centrioles]*

13. The _____ consists of four to eight flattened channels connected to the endoplasmic reticulum. The function is largely secretion of proteins, lipids, and carbohydrates. *[Golgi complex]*

14. The term ___hypotonic___ means a lower concentration of solutes outside the cell than inside.

15. Three examples of passive diffusion across membranes are ___osmosis___, ___simple diffus___, and ___facilitat diffus___.

16. Gram-___neg___ cells have a relatively smaller amount of peptidoglycan surrounded by an outer membrane of polysaccharide-phospholipid-lipoprotein than other bacterial cells.

17. The flagella of bacteria are anchored to the cell wall and plasma membrane by ___basal bodies___.

18. The protein in the flagellar filaments of bacteria is called ___flagellin___.

19. The protein forming pili is called ___pilin___.

20. Bacteria with irregular morphology are termed ___macroplasms___. *[pleomorph]*

21. Another name for extracellular polymeric substance that helps cells adhere to surfaces is ___slime layer / capsule___ *[glycocalyx]*

Key

Matching

I.   1.h 2.c 3.a 4.d 5.e 6.f 7.j
II. 1.a 2.a 3.b 4.a 5.b 6.a
III. 1.i 2.g 3.d 4.c 5.a 6.f
IV. 1.a 2.d 3.j 4.g 5.h 6.f 7.c
V.   1.f 2.g 3.e 4.d 5.b 6.c
VI. 1.d 2.a 3.c

Fill-in-the-Blanks

1. polysaccharide   2. phagocytosis   3. Plasmids   4. thylakoids   5. Pores
6. Endospores   7. septum   8. L form   9. spore septum; forespore
10. roughreticulum   11. matrix   12. centrioles   13. Golgi complex   14. hypotonic
15. simple diffusion; osmosis; facilitated diffusion   16. negative   17. basal body
18. flagellin   19. pilin   20. pleomorphic   21. glycocalyx

5 *Microbial Metabolism*

Metabolism is the sum of all chemical reactions within a living organism, including **anabolic** (synthetic) reactions and **catabolic** (degradative) reactions. Anabolism is the combination of simpler substances into complex substances and *requires* energy. Catabolism releases energy stored in organic molecules, for example, and *yields* energy. Energy liberated by catabolism is stored in energy-rich bonds of **adenosine triphosphate (ATP)**.

Enzymes

Enzymes are proteins that function as **catalysts**, substances that speed up a reaction without being changed by it; that is, they lower the **activation energy** needed. An enzyme has a three-dimensional structure, with an **active site** that reacts with the surface of the **substrate** (the substance acted upon). The **turnover number** is the number of molecules metabolized per enzyme molecule per second. Enzymes generally are named after the substrate they react with or the type of reaction they perform, using the suffix *-ase*, as in cellulase, for the substrate cellu*lose*.

Enzyme Components

Many enzymes contain an **apoenzyme**, a protein that is inactive without a **cofactor**, a non-protein component. Together, they are an activated **holoenzyme** (whole enzyme). Cofactors may serve as a bridge binding enzyme and substrate together. Cofactors may be metal ions or organic molecules called **coenzymes**. Many coenzymes are derived from vitamins; two of the most important are **NAD** (**nicotinamide adenine dinucleotide**) and **NADP** (**nicotinamide adenine dinucleotide phosphate**). Both are derivatives of the B vitamin nicotinic acid (niacin). They function as dehydrogenases, removing and transferring hydrogen ions and electrons. The flavin coenzymes, **FMN** (**flavin mononucleotide**) and **FAD** (**flavin adenine dinucleotide**) are derivatives of the B vitamin riboflavin. They also are dehydrogenases. **Coenzyme A** (CoA) is a derivative of the B vitamin pantothenic acid. It is important in reactions of the Krebs cycle, decarboxylation, and the synthesis and breakdown of fats. When tightly bonded to apoenzymes, coenzymes are called **prosthetic groups**; heme (iron-containing) groups of cytochrome *c* are an example. **Cytochromes** are a group of enzymes that function as electron carriers in respiration and photosynthesis.

Mechanism of Enzymatic Action

The substrate surface contacts a specific region of the enzyme molecule, the **active site**, forming an intermediate **enzyme-substrate complex**. The substrate is transformed by breakdown of the molecule or combination with another substrate molecule. The enzyme is freed, unchanged.

Factors Influencing Enzymatic Activity

Temperature. Most chemical reactions occur more rapidly as the temperature rises, but, above a certain point, **denaturation** of enzyme proteins results in a drastic decline in biological reaction rates. **Denaturation** usually involves breakage of the hydrogen bonds and similar weak bonds that hold the enzyme in its three-dimensional structure.

37

pH. Enzymes have a **pH optimum** at which activity is maximal. Extreme pH changes cause denaturation.

Substrate concentration. At high substrate concentrations, the enzyme may have its active site occupied at all times by substrate or product molecules; that is, it may be **saturated**. No further increase in substrate concentrations will have an effect on the reaction.

Inhibitors. There are two forms of enzyme inhibitors. **Competitive inhibitors** compete with the normal substrate for the active site of the enzyme. These inhibitors have a shape and chemical structure similar to the normal substrate. The action of sulfa drugs depends on competitive inhibition. **Noncompetitive inhibitors** decrease the ability of the normal substrate to combine with the enzyme. The site of a noncompetitive inhibitor's binding is an **allosteric site**. An **allosteric inhibitor** changes the shape of the active site. Other examples are chemicals that tie up metal ions (cofactors) and prevent enzymatic reactions, such as cyanide, which binds iron, and fluoride, which binds calcium or magnesium.

Feedback Inhibition

In some metabolic reactions, several steps are required. In many instances the final product can inhibit enzymatic activity at some step and prevent making of excessive end product. This is called **feedback inhibition** (or **end product inhibition**).

Energy Production Methods

Nutrient molecules have energy stored in bonds that can be concentrated into the high-energy bonds of ATP.

Oxidation-Reduction

Oxidation is the addition of oxygen or, more generally, the removal of electrons (e⁻) or hydrogen ions (H⁺). Because H⁺ are lost, most biological reactions are called dehydrogenation reactions. When a compound gains electrons or hydrogen atoms, or loses oxygen, it is **reduced**. Oxidation and reduction in a cell are always coupled—one substance is oxidized and another is reduced; thus, the reaction is an **oxidation-reduction** reaction. NAD and NADP commonly carry the hydrogen atoms in these oxidation-reduction reactions. These reactions are usually energy producing. Highly reduced compounds, such as glucose, with many hydrogen atoms, contain much potential energy.

Generation of ATP: Types of Phosphorylation

The energy from oxidation-reduction is used to form ATP. The addition of a phosphate group is called **phosphorylation**. In **oxidative phosphorylation**, electrons removed from organic compounds are transferred in sequence down an **electron transport chain** to an electron acceptor such as oxygen or another suitable compound, releasing energy in the process. The energy is used to make ATP from ADP by adding another phosphate. In **substrate-level phosphorylation**, no oxygen or other inorganic final electron acceptor is required. ATP is generated by the direct transfer of a high-energy phosphate from an intermediate metabolic compound to ADP. Another mechanism is **photophosphorylation**, which occurs in photosynthetic cells. Light liberates an electron from chlorophyll. The electron passes down an electron transport chain, forming ATP.

38

Chemiosmotic Mechanism of ATP Generation

In **chemiosmosis**, carrier molecules pump protons from one side of a membrane to the other. This establishes a proton gradient across the membrane. This gradient has potential energy called the **proton motive force**. Protons on the outside are transported in by the membrane-bound enzyme **adenosine triphosphatase** (**ATPase**). When this occurs, energy is released which is used to form ATP.

Nutritional Patterns Among Organisms

Phototrophs use light as their primary *energy source*; **chemotrophs** extract their energy from inorganic or organic chemical compounds. The principal *carbon source* of **autotrophs** is carbon dioxide; **heterotrophs** require an organic carbon source. These terms can be combined into terms that reflect the primary energy and carbon source.

Photoautotrophs. **Photoautotrophs** include photosynthetic bacteria. Green sulfur bacteria are anaerobes that use sulfur compounds or hydrogen gas to reduce carbon dioxide and form organic compounds. Light is the energy source. Most commonly, they produce sulfur from hydrogen sulfide. Purple sulfur bacteria, which are mostly anaerobes, also use sulfur compounds or hydrogen gas to reduce carbon dioxide. Neither of these photosynthetic bacteria use water to reduce carbon dioxide, as do plants, and they do *not* produce oxygen gas (they are **anoxygenic**) as a product of photosynthesis. Their photosynthetic pigment is bacterio-chlorophyll, which absorbs longer wavelengths of light than chlorophyll *a*. Cyanobacteria use chlorophyll *a* in photosynthesis. They produce oxygen gas, (**oxygenic**) just as higher plants do.

Photoheterotrophs. Green nonsulfur and purple nonsulfur bacteria are **photoheterotrophs**. They use light as an energy source but must use organic compounds instead of carbon dioxide as a carbon source. Otherwise, they are similar to the green and purple sulfur bacteria.

Chemautotrophs. *Inorganic* compounds such as hydrogen sulfide, elemental sulfur, ammonia, nitrites, hydrogen, and iron are used by **chemoautotrophs** as sources of energy. Carbon dioxide is their principal carbon source. These compounds contain energy that may be extracted by oxidative phosphorylation reactions.

Chemoheterotrophs. With **chemoheterotrophs**, the carbon source and energy source are usually the same *organic* compound—glucose, for example. Most microorganisms are chemoheterotrophs. **Saprophytes** live on dead organic matter and **parasites** derive nutrients from a living host.

Biochemical Pathways of Energy Production

A sequence of enzymatically catalyzed chemical reactions in a cell is a **biochemical pathway**. Such pathways are necessary to extract energy from organic compounds; they allow energy to be released in a controlled manner instead of in a damaging burst with a large amount of heat.

Carbohydrate Catabolism

Glycolysis

The six-carbon sugar glucose plays a central role in carbohydrate metabolism (Figure 5-1). Glucose is usually broken down to **pyruvic acid** by **glycolysis**, which is the first stage of both fermentation and respiration.

Glycolysis, also called the **Embden-Meyerhof pathway**, is a series of ten chemical reactions. The main points are that a six-carbon glucose molecule is split and forms two molecules of pyruvic acid. Two molecules of ATP were needed to start the reaction, and four molecules of ATP are formed by substrate-level phosphorylation; the net yield, therefore, is two ATP molecules. An alternate pathway is the **pentose phosphate pathway (hexose monophosphate shunt)**. It produces important intermediates (pentoses) that act as precursors in the synthesis of nucleic acids, certain

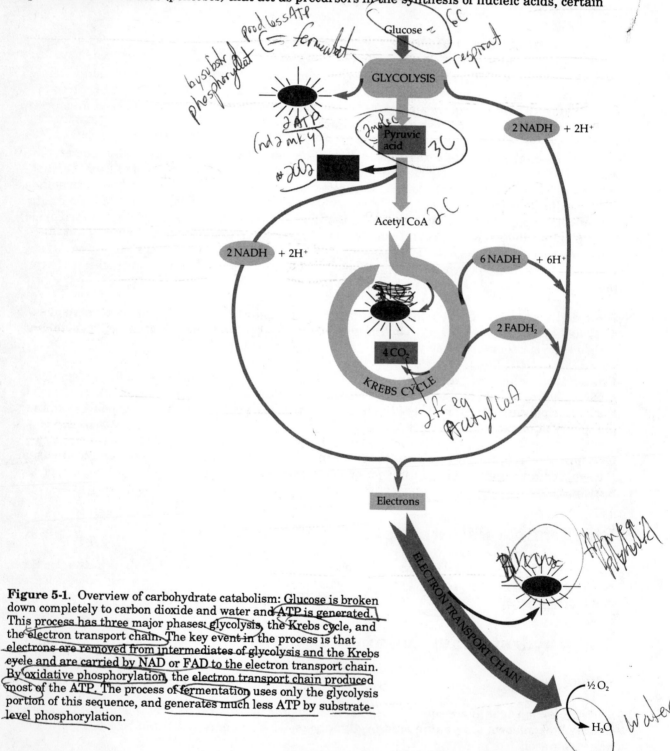

Figure 5-1. Overview of carbohydrate catabolism: Glucose is broken down completely to carbon dioxide and water and ATP is generated. This process has three major phases: glycolysis, the Krebs cycle, and the electron transport chain. The key event in the process is that electrons are removed from intermediates of glycolysis and the Krebs cycle and are carried by NAD or FAD to the electron transport chain. By oxidative phosphorylation, the electron transport chain produced most of the ATP. The process of fermentation uses only the glycolysis portion of this sequence, and generates much less ATP by substrate-level phosphorylation.

amino acids, and glucose from carbon dioxide by photosynthesizing organisms. Another way to oxidize glucose to pyruvic acid is the **Entner-Doudoroff pathway (EDP)**. Bacteria that utilize EDP can metabolize without either glycolysis or the pentose phosphate pathway. This pathway yields NADPH from glucose, which may be used for biosynthetic reactions.

Respiration

If oxygen or some other inorganic molecule is present as an electron acceptor, respiration is possible. The energy yield of respiration is much greater than it is in fermentation. When oxygen is used as an electron acceptor, it is called aerobic respiration.

Aerobic respiration. Most of the energy contained in the glucose molecule still remains after fermentation in the form of lactic acid, ethanol, etc. If the final electron acceptor is oxygen, the glucose molecule can be completely oxidized to carbon dioxide and water. Two sets of reactions beyond glycolysis must occur for aerobic respiration to take place. First, three-carbon pyruvic acid is converted to two-carbon **acetyl coenzyme A** by decarboxylation. This product can then enter the Krebs cycle. During the Krebs cycle, three molecules of carbon dioxide are liberated from each pyruvic acid molecule.

Reduced coenzymes ($NADH_2$ and $FADH_2$) are also produced from the Krebs cycle, and these now contain stored energy. To transfer these to ATP, an **electron transport chain** is required. This chain contains **flavoproteins, cytochromes**, and **ubiquinones** (or **coenzyme Q**), which are capable of oxidation and reduction. ATP is generated by a stepwise release of energy as electrons are passed along the chain. Oxygen is the final electron acceptor, where water is formed.

Anaerobic respiration. Some bacteria can use oxygen substitutes such as nitrate (forming nitrite, nitrous oxide, or nitrogen gas), sulfate (forming hydrogen sulfide), or carbonate (forming methane).

Fermentation

After glucose is broken down to pyruvic acid, the pyruvic acid can undergo **fermentation**. Fermentation does not rquire oxygen or an electron transport chain. It uses an organic molecule as the final electron acceptor. In fermentation, the pyruvic acid accepts electrons (hydrogen) and is turned into various end products such as lactic acid (Figure 5-20a in text) or ethanol (Figure 5-20b in text).

Lipid Catabolism

Some microorganisms produce extracellular enzymes, lipases, that break fats into fatty acids and glycerol. These components are then metabolized separately. Fatty acids are oxidized by **beta oxidation**, in which carbon fragments are removed two at a time to form acetyl coenzyme A, whose molecules then enter the Krebs cycle. Glycerol forms an intermediate of glycolysis and then is further oxidized.

Protein Catabolism

Proteases and peptidases are extracellular enzymes produced by some microbes to break down proteins, whose molecules are much too large for microbes to use unaltered, into component amino acids. Amino acids are first **deaminated** ($-NH_2$ removed) and **decarboxylated** ($-COOH$ removed). They then enter the Krebs cycle in various ways.

Biochemical Pathways of Energy Utilization (Anabolism)

Energy generated by catabolism may be used in the synthesis of new compounds for the cell. **Polysaccharides** such as glycogen are synthesized from glucose. Glucose is first joined with ATP, forming adenosine diphosphoglucose (ADPG). The energy of the ATP is, in essence, used to fasten together the sequence of glucose molecules to form a polysaccharide. **Lipids** such as fats are formed by combining glycerol with fatty acids. The glycerol is derived from a glycolysis intermediate, and the fatty acids are built up from two-carbon fragments of acetyl coenzyme A. The **amino acids** required for the biosynthesis of proteins are synthesized by some bacteria; other bacteria require them to be preformed. Intermediates of carbohydrate metabolism are used in the synthesis of amino acids. DNA and RNA are made up of repeating units called **nucleotides**. These consist of a purine or pyrimidine, a five-carbon sugar, and a phosphate group. Sugars for nucleotides are derived from the pentose phosphate pathway or the Entner-Doudoroff pathway. Amino acids such as glycine and glutamine furnish the atoms from which are derived the backbone of purines and pyrimidines.

Integration of Metabolism

Anabolic and catabolic reactions are integrated through common intermediates. The Krebs cycle, for example, can operate in both anabolic and catabolic reactions. Such pathways are called **amphibolic pathways**.

Test

In the matching section, there is only one answer to each question; however, the lettered options (a, b, c, etc.) may be used more than once or, perhaps, not at all.

I. Match

1. Energy-yielding series of reactions. _a._   a. Catabolism
                                                      b. Anabolism
2. Means "whole enzyme". _f_   c. Turnover number
                                                      d. Apoenzyme
3. A nonprotein component of an active enzyme. _e_   e. Coenzyme
                                                      f. Holoenzyme
4. A measure of the rate of activity of an enzyme. _c_   g. Beta oxidation
                                                      h. Cytochrome
5. A protein portion of an enzyme, inactive without a cofactor. _d_

6. A group of enzymes that function as electron carriers _h_
   in respiration and photosynthesis.

7. A mechanism by which fatty acids are degraded. _g_

II. Match

1. Both the carbon source and energy source _g_   a. Competitive inhibitor
   are usually the same organic compound.
                                                         b. Allosteric inhibitor
2. Photosynthetic, but uses organic material rather _f_
   than carbon dioxide as a carbon source.   c. Noncompetitive inhibitor

3. The photosynthetic purple nonsulfur bacteria _f_   d. Photoautotroph
   would be classified in this nutritional group.
                                                         e. Chemoautotroph
4. Photosynthetic bacteria that do not produce
   oxygen gas, and use carbon dioxide as a carbon source. _d_   f. Photoheterotroph

5. Changes the shape of the active site of the enzyme. _b_   g. Chemoheterotroph

6. Very similar in shape or chemistry to the normal _a_
   enzyme substrate.

43

III. Match

1. Hexose monophosphate shunt. *c*
2. Final electron acceptor is oxygen. *f*
3. Yields only one ATP; produces important intermediates that act as precursors in the synthesis of nucleic acids, etc. *c*
4. Bacteria use oxygen substitutes such as nitrates. *e*
5. Pyruvic acid accepts electrons and is turned into various end products such as lactic acid or ethanol. *a*
6. Glucose to pyruvic acid. *b*

a. Fermentation
b. Glycolysis
c. Pentose phosphate pathway
d. Substrate-level phosphorylation
e. Anaerobic respiration
f. Aerobic respiration

IV. Match

1. Electrons are removed from an organic compound and transferred by electron transport chain to oxygen. *a*
2. An electron is liberated from chlorophyll and passes down an electron transport chain *c*

a. Oxidative phosphorylation
b. Substrate-level phosphorylation
c. Photophosphorylation

V. Match

1. A dehydrogenase coenzyme derived from nicotinic acid (niacin). *a*
2. A dehydrogenase coenzyme derived from riboflavin. *d*
3. An enzyme that transports protons across membrane. *e*
4. Coenzyme Q. *b*

a. NAD
b. Ubiquinone
c. Coenzyme A
d. FMN
e. ATPase

VI. Fill-in-the-Blanks

1. A chemoheterotroph that lives on dead organic matter is called a _Saprophyte_.
2. When tightly bonded to apoenzymes, coenzymes are called _holoenzyme prosthetic grps_
3. When an enzyme's active site occupied at all times by substrate or product molecules, it is _saturated_
4. Cyanide is an example of what general type of inhibitor? _competitive allosteric noncompetitive_

5. Sulfa drugs are an example of what type of inhibitor? _competv_

6. The removal of electrons or hydrogen ions, or the addition of oxygen, is termed _Reduction oxidat_

7. In _substrate l_ phosphorylation, no oxygen or other inorganic final electron acceptor is required.

8. Cyanobacteria produce _O_2_ gas, just as do higher plants.

9. The amount of ATP yield from aerobic respiration is _38 m_.

10. The amount of ATP yield from glycolysis is _2 ATP_.

11. The removal of NH_2 from an amino acid is called _deaminat_.

12. The removal of COOH from an amino acid is called _decarboxylat_.

13. The substance acted upon by an enzyme is called the _substrate_.

14. Coenzyme A is a derivative of the B vitamin _niacin pantothenic A_

15. Inorganic sources of energy such as hydrogen sulfide, elemental sulfur, ammonia, nitrites, etc., are used by the _heterotroph chemo autotroph_ nutritional group of bacteria as a source of energy.

16. A sequence of enzymatically catalyzed chemical reactions in a cell is called a _Biochem pathwy_

17. Glucose is usually broken down to pyruvic acid by _glycolysis_.

18. Another name for glycolysis is the _Embden-Meyerhof pthwy_.

19. In aerobic respiration, pyruvic acid is converted to acetyl _CoA_; this product can then enter the Krebs cycle.

20. DNA and RNA are made up of repeating units called _nucleotides_.

Key

Matching

I.   1.a   2.f   3.e   4.c   5.d   6.h   7.g
II.   1.g   2.f   3.f   4.d   5.b   6.a
III. 1.c   2.f   3.c   4.e   5.a   6.b
IV. 1.a   2.c
V.   1.a   2.d   3.e   4.b

Fill-in-the-Blanks

1. saprophyte   2. prosthetic groups   3. saturated   4. noncompetitive   5. competitive
6. oxidation   7. substrate-level   8. oxygen   9. 38   10. 2   11. deamination
12. decarboxylation   13. substrate   14. pantothenic acid   15. chemoautotroph
16. biochemical pathway   17. glycolysis   18. Embden-Meyerhof pathway   19. coenzyme A
20. nucleotides

6 *Microbial Growth*

Microbial growth refers to the number of cells, not to the changes in the size of cells.

Requirements for Growth

Physical Requirements

Temperature. The *minimum* growth temperature is the lowest temperature at which the species will grow, and the *maximum* is the highest. The *optimum* growth temperature is that at which it grows best.

Psychrophiles are organisms capable of growth at 0°C. Some define psychrophiles more narrowly as having an optimum growth temperature of about 15 to 20°C, and a range between 0 and 30°C. These usually are found in oceans or arctic environments. The term **psychrotrophs** has been proposed for organisms that grow well at refrigerator temperatures but which have an optimum in the upper teens or low twenties. The maximum temperature for psychrotrophs may be below 30°C. **Mesophiles** are the most common microbes; their optimum temperatures are 25 to 40°C. **Thermophiles** are capable of growth at high temperatures. Many have an optimum of 50 to 60°C. Many are not capable of growth below about 45°C.

pH. Most bacteria grow best in a narrow pH range near neutrality, between pH 6.5 and 7.5. Very few grow below pH 4.0 However, many bacteria, such as the **acidophiles** responsible for acid fermentations, are remarkably tolerant of acidity. Alkalinity inhibits microbial growth but is rarely used in food preservation. A **buffer** is sometimes added to media to neutralize acids. Examples of buffers are phosphate salts, peptones, and amino acids.

Osmotic pressure. If the microbial cell is in a solution in which the concentration of solutes is higher than that found in the cell, cellular water passes through the cytoplasmic membrane in the direction of the high salt concentration. The cytoplasmic membrane collapses away from the cell wall during the loss of water, which is called **plasmolysis**. **Extreme halophiles** are organisms that have so adapted to high salt concentrations that they require them for growth. **Facultative halophiles** do not require high salt concentrations, but they are able to grow at salt concentrations that tend to inhibit the growth of many other bacteria. The salt concentration may be 1 or 2% or as high as 10 to 15%. Osmotic effects are roughly related to the numbers of molecules in a given volume of solution. For a given weight/volume, therefore, sodium chloride is more effective than sucrose.

Chemical Requirements

Carbon. Besides water, which is needed because nutrients must be in solution in order to be used, carbon is a primary requirement for growth. It is the structural backbone of living matter. Its four valences allow it to be used in constructing complicated organic molecules. Sources of carbon are carbon dioxide or organic materials.

Nitrogen, sulfur, and phosphorus. Some organisms use proteinaceous material as a nitrogen source; others use ammonium ions (NH_4^+) or nitrate ions (NO_3^-). A few bacteria and *Cyanobacteria* are able to use gaseous nitrogen (N_2) directly from the atmosphere. This process is called **nitrogen**

fixation. The *Rhizobium* and *Bradyrhizobium* bacteria, in symbiosis with leguminous plants, also fix nitrogen. Sulfur sources are sulfate ion (SO_4^{2-}), hydrogen sulfide (H_2S), and the sulfur-containing amino acids. An important source of phosphorus is the phosphate ion (PO_4^{3-}). Nitrogen, sulfur, and proteins are required to synthesize proteins—for example, DNA, RNA, and ATP.

Trace elements. Mineral elements such as iron, copper, and zinc are referred to as **trace elements**. Although sometimes added to laboratory media, they usually are assumed to be naturally present in water and other media components.

Oxygen. Microbes that use molecular oxygen are **aerobes**; if oxygen is an absolute requirement, they are **obligate aerobes**. **Facultative anaerobes** use oxygen when it is present, but continue growth by fermentation or anaerobic respiration when it is not available. Facultative anaerobes grow more efficiently aerobically than they do anaerobically. **Obligate anaerobes** are bacteria totally unable to use oxygen for growth and usually find it toxic. Hydrogen atoms in the electron transport chain may be passed to oxygen, forming toxic hydrogen peroxide (H_2O_2). Aerobic organisms usually produce **catalase**, an enzyme that breaks down hydrogen peroxide to water and oxygen; anaerobes usually lack catalase. **Aerotolerant anaerobes** cannot use oxygen for growth but tolerate it fairly well. They will grow on the surface of a solid medium without the special techniques required for cultivation of less oxygen-tolerant anaerobes. Common examples of aerotolerant anaerobes are the bacteria that ferment carbohydrates to lactic acid, which occurs in making many fermented foods. A few bacteria are **microaerophilic**, meaning they grow only in oxygen concentrations lower than found in air. They are aerobic, however, in the sense that they require oxygen. They are probably unusually sensitive to superoxide free radicals and peroxides.

Oxygen has a number of toxic forms.

1. **Singlet oxygen** is normal molecular oxygen (O_2) that has been boosted into a higher energy state and is extremely reactive.

2. **Superoxide free radicals** (O_2^-) are formed in small amounts by aerobic organisms; they are so toxic that the bacteria must neutralize them with **superoxide dismutase**. This enzyme combines oxygen into hydrogen peroxide, which in turn is converted into oxygen and water by the enzyme **catalase**: $2H_2O_2 \longrightarrow 2H_2O + O_2$. Another enzyme that breaks down hydrogen peroxide is **peroxidase**; it does not produce O_2: $H_2O_2 + 2H^+ \longrightarrow 2H_2O$. Anaerobic bacteria often cannot neutralize the superoxide free radicals they produce and do not tolerate atmospheric oxygen.

3. **Peroxide** (O_2^{2-}) is the active principle in antimicrobials such as hydrogen peroxide and benzoyl peroxide.

4. The hydroxyl free radical (OH·) is formed in cytoplasm by ionizing radiation and as a byproduct of aerobic respiration. It is probably the most reactive form.

Organic growth factors. **Organic growth factors** are organic compounds such as vitamins, amino acids, and pyrimidines that are needed for life, but that a given organism is unable to synthesize.

Cultural Media

Any nutrient material prepared for the growth of bacteria in a laboratory is called a **culture medium**. Microbes growing in a container of culture medium are referred to as a **culture**. To ensure that the culture will contain only the microorganisms originally added to the medium (and their offspring), the medium must initially be sterile. When a solid medium is required, a solidifying agent such as **agar** is added. Agar is a polysaccharide derived from a marine alga. Few microbes can degrade agar, so it remains a solid. It melts at about the boiling point of water but remains liquid until the temperature drops to about 40°C.

Chemically Defined Media

In a **chemically defined medium**, the exact chemical composition is known.

Complex Media

Most heterotrophic bacteria and fungi are routinely grown on **complex media**, in which the exact chemical composition varies slightly from batch to batch. Complex media are made up of nutrients such as extracts from yeasts, beef, or plants, or digest of proteins from these and other sources. In many of these media, the energy, carbon, nitrogen, and sulfur requirements of the microorganisms are largely met by partially digested protein products called **peptones**. Vitamins and other organic growth factors are provided by meat extracts or yeast extracts. Such extracts supplement the organic nitrogen and carbon compounds but mainly are sources of soluble vitamins and minerals. This type of medium in liquid form is **nutrient broth**; when agar is added, it is **nutrient agar**.

Anaerobic Growth Media and Methods

Obligately anaerobic bacteria often require reducing media for isolation. Because oxygen may be lethal, these media contain ingredients such as sodium thioglycolate that chemically combine with dissolved oxygen to deplete the oxygen content of the culture medium. Obligate anaerobes may be grown on the surface of solid media in anaerobic atmospheres produced in special jars in which an oxygen-free atmosphere is generated by a chemical reaction. Surface growth also can be accomplished by substituting deep vials for Petri dishes. In this process, an inert gas replaces the atmosphere in the open vial, and then the bacterial inoculum is mixed with melted nutrient medium and pipetted into the vial. The vial is tightly capped and rolled on horizontal rollers to distribute the medium and bacteria against the interior walls. Colonies appear in this film of medium and are easily counted. It is also possible to handle anaerobic organisms in anaerobic glove boxes fitted with airtight rubber glove arms and air locks, and filled with inert gases.

Special Culture Techniques

A few bacteria have never been successfully grown on artificial laboratory media—the leprosy and syphilis organisms are examples. Obligate intracellular parasites such as rickettsia and chlamydia ordinarily also do not grow on artificial media. They, like viruses, require a living host cell. Carbon dioxide incubators and candle jars are used to grow bacteria with special carbon dioxide concentration requirements.

Selective and Differential Media

Selective media are designed to suppress the growth of unwanted bacteria and encourage the growth of the desired microorganisms. Antibiotics, high concentrations of salt, or high acidity might be used. **Differential media** make it easier to distinguish colonies of the desired organism from other colonies growing on the same plate. The colonies have different colors or cause different changes in the surrounding medium. Sometimes selective and differential functions are combined in the one medium.

Enrichment culture. Because bacteria may be present only in small numbers and may be missed, and because the bacterium to be isolated may be of an unusual physiological type, it is sometimes necessary to resort to enrichment cultures. For example, only a tiny number of organisms in a soil sample are capable of growing on phenol, a disinfectant. These organisms can be isolated, however, by placing the soil sample in a medium in which the only source of carbon and energy is phenol. A series of transfers between fresh culture media of the same type, each of which is allowed to incubate for a few days, gradually eliminates organic matter from the original inoculum. Finally, only organisms capable of growth on phenol will survive.

Obtaining Pure Cultures

There are several methods for isolating bacteria in **pure culture** that contain only one kind of organism.

Streak plate method. Probably the most common method of obtaining pure cultures is the **streak plate** (Figure 6-1). A sterile inoculating needle is dipped into a mixed culture and streaked in a pattern over the surface of the nutrient medium. The last cells that are rubbed from the needle are wide enough apart that they grow into isolated visible masses called **colonies**.

Preserving Bacterial Cultures

In **deep-freezing**, a pure culture of microbes is placed in a suspending liquid and quick frozen at -50 to -95°C. In **lyophilization (freeze-drying)**, a suspension of microbes is quickly frozen at temperatures from -54 to -72°C and the water removed by a high vacuum. The remaining powder can be stored for many years and the surviving microorganisms cultured by hydrating them with a suitable liquid nutrient medium.

Growth of Bacteria Cultures

Bacterial Division

Bacteria normally reproduce by **binary fission**. Genetic material becomes evenly distributed; then a transverse wall is formed across the center of the cell and it separates into two cells. A few bacterial

Figure 6-1. Streak plate method for isolation of pure cultures of bacteria. The direction of streaking is indicated by arrows. Streak series I is made from the original bacterial mixture. The inoculating loop is sterilized following such streak series, which dilutes out the number of cells in each succeeding series. In this way, the bacteria are diluted out, and well-isolated colonies are obtained

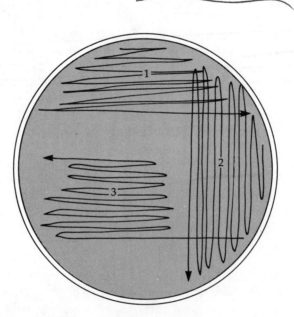

species **bud**; that is, an initial outgrowth enlarges to cell size and then separates. Some filamentous species produce **reproductive spores** or simply **fragment** into viable fragments.

Generation Time

The time required for the cell to divide or the population to double is called **generation (doubling) time**. Bacterial populations are usually graphed **logarithmically** rather than **arithmetically** to permit the handling of the immense differences in numbers (Figure 6-2).

Phases of Growth

When bacterial population changes are graphed as a bacterial growth curve, certain phases become apparent. The **lag phase** shows little or no cell division. However, metabolic activity is intense. In the **log phase**, the cells are reproducing most actively, and their generation time reaches a minimum and remains constant; a logarithmic plot produces an ascending straight line. They are then most active metabolically and most sensitive to adverse conditions. In a **chemostat**, it is possible to keep a population in such exponential growth indefinitely. Without a chemostat, however, microbial deaths eventually balance numbers of new cells and a **stationary phase** is reached. When the number of deaths exceeds numbers of new cells formed, the **death phase**, or **logarithmic decline**, is reached. The entire culture may die out in time.

Measurement of Microbial Growth

Plate counts. Dilutions of a bacterial suspension are distributed into a suitable solid nutrient medium by **serial dilution**, and the colonies appearing on the plates are counted (Figure 6-3).

Filtration. Bacteria may be sieved out of a liquid suspension onto a thin membrane filter with pores too small for bacteria to pass. This filter can be transferred to a pad soaked in nutrient medium where colonies arise on the surface of the filter.

Most probable number (MPN). In the most probable number method, a sample is diluted out in a series of tubes of liquid medium. The greater the number of bacteria, the more dilutions it takes to dilute them out entirely and leave a tube without growth. Results of such dilutions can be compared to statistical tables, and a cell count can be estimated.

Figure 6-2. Growth curve for an exponentially increasing bacterial population plotted logarithmically and arithmetically

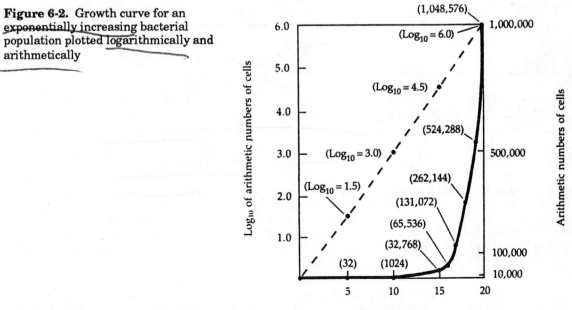

51

Direct microscopic count. In this method, a measured volume of a bacterial suspension is placed into a definite area on a special microscope slide, of which there are several designs. A microscope is used to count the cells in microscope fields. The average number per field can be multiplied by a factor that will estimate the total numbers.

Turbidity. To estimate turbidity, a beam of light is transmitted through a bacterial suspension to a photoelectric cell. The more bacteria, the less light passes. This is recorded as **absorbance** (sometimes *Optical Density* or *OD*) on the spectrophotometer or colorimeter.

Metabolic activity. Microbial numbers can be estimated by the time required to deplete oxygen (**reduction tests**) or to produce acid or other products.

Dry weight. Measuring weight often is the most satisfactory method for filamentous organisms such as fungi.

Figure 6-3. (Opposite) Plate counts and serial dilutions. (a) In serial dilutions, the original inoculum is diluted out in a series of dilution tubes. In our example, each succeeding dilution tube will have only one tenth of the number of microbial cells as the preceding tube. **(b)** Pour plate. In making a pour plate, the inoculum is placed in a Petri plate. The melted agar medium is poured over it and allowed to solidify. **(c)** Spread plate. In making a spread plate, 0.1 ml of inoculum is placed on the surface of a prepoured plate of solidified agar medium and then is spread evenly.

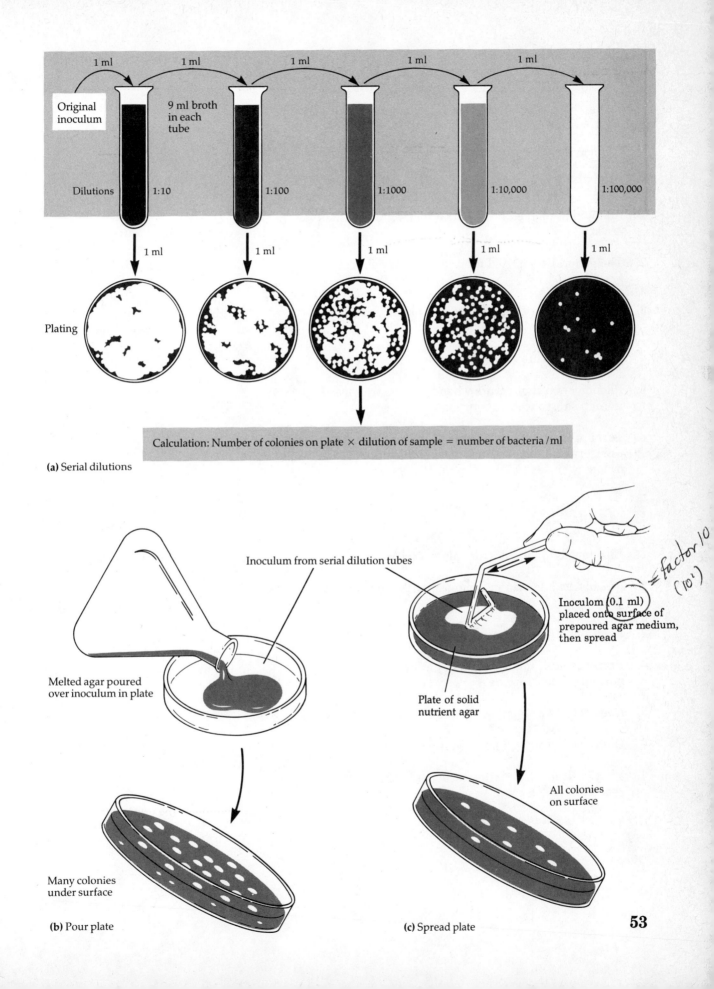

(a) Serial dilutions

Dilutions: 1:10 | 1:100 | 1:1000 | 1:10,000 | 1:100,000

1 ml | 1 ml | 1 ml | 1 ml | 1 ml

Original inoculum

9 ml broth in each tube

Plating

Calculation: Number of colonies on plate × dilution of sample = number of bacteria / ml

Inoculum from serial dilution tubes

Melted agar poured over inoculum in plate

Many colonies under surface

(b) Pour plate

Inoculom (0.1 ml) placed onto surface of prepoured agar medium, then spread

Plate of solid nutrient agar

= factor 10 (10^1)

All colonies on surface

(c) Spread plate

53

Test

In the matching section, there is only one answer to each question; however, the lettered options (a, b, c, etc.) may be used more than once or, perhaps, not at all.

I. Match

1. Adapted to high salt concentrations; requires them for growth. *g*

2. The general term used for organisms capable of growth at 0°C. *d*

3. Capable of growth at high temperatures; optimum 50 to 60°C. *c*

4. Used in media to neutralize acids. *a*

5. Phenomenon that occurs when bacteria are placed in high salt concentration. *f*

6. Term proposed for organisms that grow well at refrigerator temperatures; optimum is upper teens or low twenties. *e*

a. Buffer
b. Mesophile
c. Thermophile
d. Psychrophile
e. Psychrotroph
f. Plasmolysis
g. Extreme halophile
h. Facultative halophile

II. Match

1. Enzyme acting upon hydrogen peroxide. *e*

2. *Rhizobium* bacteria in symbiosis with leguminous plants. *a*

3. Requires atmospheric oxygen to grow. *b*

4. Requires atmospheric oxygen, but in lower than normal concentrations. *f*

5. Does not use oxygen, but grows readily in its presence. *d*

6. Does not use oxygen and usually finds it toxic. *c*

7. Important for the energy, carbon, nitrogen, and sulfur requirements in complex media. *g*

a. Nitrogen fixation
b. Obligate aerobe
c. Obligate anaerobe
d. Aerotolerant anaerobe
e. Catalase
f. Microaerophile
g. Peptones
h. Agar

III. Match

1. Isolation method for getting pure cultures; uses an inoculating loop to trace a pattern of inoculum on a solid medium. — b

2. A device for maintaining bacteria in a logarithmic growth phase. — i

3. Used to increase the numbers of a small minority of microorganisms in a mixed culture to arrive at a pure culture. — ~~f~~ ~~f~~ f

4. Preservation method that uses quick-freezing and a high vacuum. — g

a. Pour plate
b. Streak plate
c. Spread plate
d. Differential medium
e. Reducing medium
f. Enrichment culture   (? select)
g. Lyophilization
h. Deep-freezing
i. Chemostat

IV. Match

1. New cell numbers balanced by death of cells. — d

2. No division but intense metabolic activity. — b

3. A logarithmic plot of the population produces an ascending straight line. — a

a. Log phase
b. Lag phase
c. Death phase
d. Stationary phase

V. Match

1. Used to grow obligate anaerobes. — d

2. Designed to suppress the growth of unwanted bacteria and to encourage growth of desired microbes. — ~~b~~ a

3. Generally contain ingredients such as sodium thioglycolate, which chemically combine with. dissolved oxygen — ~~b~~ d

4. Nutrients are digests or extracts; exact chemical composition varies slightly from batch to batch. — c

a. Selective media

b. Differential media

c. Complex media

d. Reducing media.

e. Chemically defined media

VI. Match

1. Active principle in benzoyl peroxide. ~~a~~ b   a. Hydroxyl free radical
                                          b. Peroxide

2. Formed in cytoplasm by ionizing radiation. ~~a~~ ~~e~~ a   c. Superoxide dismutase
                                          d. Superoxide free radicals

3. An enzyme that converts hydrogen peroxide into f   e. Singlet oxygen
oxygen and water.   f. Catalase

4. The toxic form of oxygen neutralized by superoxide d
dismutase.

VII. Fill-in-the-Blanks

1. Agar is a ___Polysaccharide___ derived from a marine alga.

2. A few bacteria and the ___legumes cyanobacter___ are able to use gaseous nitrogen directly from the atmosphere.

3. ___Mesophiles___ are the most common microbes; their optimum temperatures are 25-40°C.

4. Osmotic effects are roughly related to the _____~~☐~~ #_____ of molecules in a given volume of solution.

5. A complex medium in liquid form is called nutrient ___broth___.

6. For preservation by ___deep freez___, a pure culture of microbes is placed in a suspending liquid and quick frozen at -50 to -95°C.

7. Bacteria usually reproduce by ___binary___ fission.

8. Turbidity is recorded in a spectrophotometer as am't ___absorpt / (OD) optic density___

9. The growth of filamentous organisms such as fungi is often best recorded by means of ___dry w measurm___

10. ___Facultatv___ anaerobes grow more efficiently aerobically than they do anaerobically.

11. ___facultatv___ halophiles do not require high salt concentrations, but they are able to grow at salt concentrations that may inhibit the growth of many other bacteria.

12. Examples of buffers are ___phosphate___ salts; peptones and ___amino A's___ found in complex media are also buffers.

13. Any nutrient material prepared for the growth of bacteria in a laboratory is called a ___cult medium___.

14. Agar melts at about the boiling point of water but remains liquid until the temperature drops to about ___40°C___.

15. Dilutions of a bacterial mixture are poured into a Petri dish and mixed with melted agar. . This plate counting method is called the ___*pour plate*___ .

16. Partially digested protein products used in complex media are called ___*nutrients*___ . *peptones*

17. In order to grow obligate intracellular parasites such as rickettsia and chlamydia, it is usually necessary to provide ___*living host cells*___

18. The general term for tests that estimate microbial growth by the time required for them to deplete oxygen in the medium are called ___*metabolic*___ tests. *reduction*

19. The ___*optimum*___ growth temperature is that at which the organism grows best.

Key

Matching

I. 1.g 2.d 3.c 4.a 5.f 6.e
II. 1.e 2.a 3.b 4.f 5.d 6.c 7.g
III. 1.b 2.i 3.f 4.g
IV. 1.d 2.b 3.a
V. 1.d 2.a 3.d 4.c
VI. 1.b 2.a 3.f 4.d

Fill-in-the-Blanks

1. polysaccharide   2. *Cyanobacteria*   3. mesophiles   4. number   5. broth
6. deep-freezing   7. binary   8. absorbance (also Optical Density)   9. dry weight measurement
10. facultative   11. facultative   12. phosphate; amino acids   13. culture medium   14. 40°C
15. pour plate method   16. peptones   17. living host cells   18. reduction   19. optimum

7 *Control of Microbial Growth*

Terminology Related to the Control of Microbial Growth

Sterilization is the destruction of all forms of microbial life. **Disinfection** is the destruction of vegetative pathogens on a surface, usually with chemicals. Spores and viruses are not necessarily destroyed. **Antisepsis** is the chemical disinfection of living tissue such as skin or mucous membrane. **Asepsis** is the absence of pathogens on an object or area, as in aseptic techniques. **Degerming** is the removal of transient microbes from skin by mechanical cleansing or by an antiseptic. **Sanitization** is the reduction of microbial populations on objects to safe public health levels. In general, the suffix *-cide* indicates the killer of a specified organism, as in **germicide, fungicide, sporicide, virucide**, etc. The suffix *-stat* used in this way indicates only that the substance inhibits, not kills.

Conditions Influencing Microbial Control

Control of microorganisms can be affected by the **temperature** of chemical solutions (generally disinfectants are more efficient when warm), by the **type of microbe** involved (pseudomonads, tuberculosis bacillus, endospores, protozoan cysts, and many viruses are relatively resistant). For example, pseudomonad resistance is related to the small size of their **porins** (wall openings). The **physiological state of the organism** (actively growing organisms are more susceptible), the **environment** (organic matter such as feces or vomit may inactivate disinfectants).

Actions of Microbial Control Agents

Alteration of Membrane Permeability

The plasma membrane controls the passage of nutrients and wastes into and out of the cell. Damage to the plasma membrane causes leakage of cellular contents and interferes with cell growth.

Damage to Proteins and Nucleic Acids

Chemicals may denature proteins by reacting, for example, with disulfide bonds, which give proteins their three-dimensional active shape. Chemicals and radiation may prevent proper replication or functioning of DNA or RNA.

Rate of Microbial Death

Bacterial populations killed by heat or chemicals tend to die at constant rates—for example, 90% every 10 minutes. Plotted logarithmically, these figures form straight descending lines.

58

Physical Methods of Microbial Control

Heat

Thermal death point is the lowest temperature required to kill a liquid culture of a certain species of bacteria in 10 minutes at pH 7. **Thermal death time** is the length of time required to kill all bacteria in a liquid culture at a given temperature. **Decimal reduction time** is that time, in minutes, required to kill 90% of the population of bacteria at a given temperature.

Moist heat sterilization. Boiling (100°C) kills vegetative forms of bacterial pathogens, many viruses, and fungi within 10 minutes. Endospores and some viruses survive boiling for longer times. **Steam under pressure** allows temperatures above boiling to be reached. **Autoclaves,** retorts, and pressure cookers are vessels in which high steam pressures can be contained. Fifteen pounds per square inch (121°C) for 15 minutes is a typical operating condition for sterilization. Moisture must touch all surfaces in order to bring about sterilization. Air must be completely exhausted from the container. An autoclave is shown in Figure 7-1.

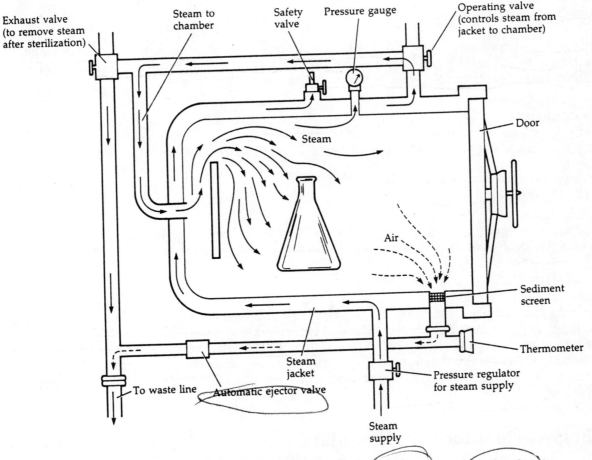

Figure 7-1. Autoclave and its path of steam flow (solid arrows). The autoclave is a steam chamber; it withstands more than 1 atmosphere of pressure (15 psi). As steam enters the chamber, air is forced out through a vent in the bottom (broken arrows). The automatic ejector valve stays open while an air-steam mixture is passing out of the waste line. When all the air has been ejected, the higher temperature of the remaining steam closes the valve. As steam continues to enter, pressure in the chamber increases.

Pasteurization is mild heating that is sufficient to kill particular spoilage or disease organisms without seriously damaging the taste of the product. **High-temperature, short-time pasteurization** uses temperatures of at least 72°C for about 15 seconds to pasteurize milk. **Equivalent treatments** are illustrated by, for example, heat of 115°C acting on an organism for 70 minutes as equivalent to heat of 125°C acting on an organism for only 7 minutes. That is, applying a higher temperature for a shorter time may kill the same number of microbes as a lower temperature for a longer time.

Dry heat sterilization. Incineration, as in direct flaming, is efficient for limited purposes. **Hot-air sterilization,** as in an oven, requires higher temperatures (such as 170°C) and longer times (such as 2 hours) to ensure sterilization. Moist heat is generally more efficient.

Filtration

Liquids sensitive to heat can be passed through a thin **membrane filter** of cellulose esters or plastic polymers that has carefully controlled pore sizes to retain microorganisms. Operating theatre and special clean rooms receive air passed through **high efficiency particulate air filters.**

Low Temperature

Refrigerator temperatures (0 to 7°C) slow the metabolic rate of microbes; however, psychrophilic species still grow slowly. Some organisms grow at temperatures slightly below freezing, but microbes at the usual temperatures of freezer compartments are completely dormant.

Desiccation

Microbes require water for growth, and adequately dried (**desiccated**) foods will not support their growth.

Osmotic Pressure

High salt or sugar concentrations cause water to leave the cell; this is an example of **osmosis** (see plasmolysis in Chapter 6). Generally, molds and yeasts resist osmotic pressures better than bacteria.

Radiation

Ionizing radiation such as x-rays, gamma rays, and high energy electron beams carry high energy and break DNA strands. Such radiation is used to sterilize pharmaceuticals. **Nonionizing radiation** such as ultraviolet light (UV) has a longer wavelength and less energy. UV causes bonds to form between adjacent thymines in DNA chains. The most effective wavelength of UV is about 260 nm. Penetration is low. Sunlight has some germicidal activity, mainly due to formation of singlet oxygen (see Chapter 6) in the cytoplasm.

Chemical Methods of Microbial Control

Evaluating a Disinfectant

Phenol coefficient. The relative effectiveness of disinfectants is determined by tests such as the **phenol coefficient test.** *Staphylococcus aureus, Pseudomonas aeruginosa,* and *Salmonella typhi* bacteria are tested under standard conditions against concentrations of phenol and the test disinfectant. If the test disinfectant is as efficient as phenol, but at a lower concentration, it has a phenol coefficient greater than 1.

Use-dilution test. The **use-dilution test** is more elaborate. A series of tubes containing increasing concentrations of the test disinfectants are inoculated and incubated. The more the chemical can be diluted and still be effective, the higher its rating.

Filter paper method. A disk of filter paper is soaked in a chemical agent, which is placed on an inoculated surface of an agar plate. A clear zone around the disk indicates inhibition.

Kinds of Disinfectants

Phenol and phenolics. Phenol (carbolic acid) is seldom used today. Derivatives of the phenol molecule, however, are widely used. **Phenolics** injure plasma membranes, inactivate enzymes, or denature proteins. They are stable, persistent, and are not sensitive to organic matter. **O-phenylphenol**, a cresol, is the main ingredient in most formulations of Lysol®. **Hexachlorophene** is effective against staphylococcal and streptococcal bacteria that cause skin infections. Problems of neurological damage to infants from excessive use have occurred, however.

Chlorhexidine. **Chlorhexidine** is not a phenol but its structure and applications resemble hexachlorophene. It is frequently used for surgical skin preparation.

Halogens

Iodine is effective against all kinds of bacteria, many endospores, fungi, and some viruses. Its mechanism of activity may be its combination with the amino acid tyrosine in enzyme and cellular proteins. A tincture of iodine means in an aqueous-alcohol solution. An **iodophore** is a combination of iodine and an organic molecule. Iodophores do not stain and are less irritating than iodine. Examples are the disinfectant Wescodyne and the antiseptic Betadine. **Chlorine** is used as a gas or in combination with other chemicals. Chlorine gas is used for disinfecting municipal water supplies, swimming pools, and sewage. Sodium hypochlorite, ordinary household bleach, is a good disinfectant. Chloramines consist of chlorine and ammonia. They are more stable than most chlorines. The germicidal action of all chlorines is based on the formation of hypochlorous acid in water. One factor in its effectiveness is its neutral charge, which allows it to diffuse as rapidly as water into the cell.

Alcohols. Both **ethanol** and **isopropanol** (rubbing alcohol) are widely used, normally at a concentration of about 70%. Concentrations of 60 to 95% are effective. They are bacteriocidal and fungicidal, but are not effective against endospores or nonenveloped viruses. Alcohols enhance the effectiveness of other chemical agents.

Heavy metals and their compounds. The fact that tiny amounts of heavy metals are effective antimicrobials can be illustrated by oligodynamic action. A silver coin on an inoculated nutrient medium will inhibit growth for some distance. A 1% silver nitrate solution is used to prevent gonorrheal eye infections in the newborn. **Silver** combines with sulfhydryl groups on proteins, denaturing them. Mercuric chloride and mercuric oxide are highly bacteriocidal, but they are toxic and corrosive and are inactivated by organic matter. **Organic mercury compounds** such as Mercurochrome and Merthiolate are less irritating and less toxic than inorganic mercuries. **Copper sulfate** often is used to destroy green algae in reservoirs or other waters. **Zinc chloride** is used in mouthwashes; and **zinc oxide** is used in paints as an antifungal.

Surface-active agents. **Surface-active agents**, or surfactants, decrease the surface tension of a liquid. Soaps and detergents are examples. They emulsify oils and are good degerming agents. **Acid-**

Figure 7-2. The ammonium ion and a quaternary ammonium compound. Note how other groups replace the hydrogens of the ammonium ion, benzalkonium chloride (Zephiran™).

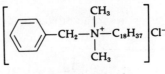

Ammonium ion

Benzalkonium chloride

anionic sanitizers are important for cleaning dairy equipment. Deodorant soaps contain compounds such as triclocarbon that inhibit gram-positive bacteria.

Quaternary ammonium compounds. The **quaternary ammonium compounds** (Figure 7-2) are most effective against gram-positive bacteria, and less so against gram-negative bacteria. These *cationic* detergents have good fungicidal, amoebicidal, and virucidal (enveloped viruses) activity, but they are not sporicidal. They are colorless, odorless, tasteless, nontoxic, and stable, but they are inactivated by organic matter, soaps, detergents, and surfaces such as gauze. They may even support the growth of *Pseudomonas* bacteria. They act by disrupting the plasma membranes and denaturing enzymes. Widely used examples of quats are benzalkonium chloride (Zephiran®), and cetylpyridinium chloride (Cepacol®).

Organic acids and derivatives. **Sorbic acid** (potassium sorbate) inhibits mold spoilage in foods such as cheese. **Benzoic acid** (sodium benzoate) is an antifungal used in soft drinks and other acidic foods. **Methylparaben** and **propylparaben**, which are derivatives of benzoic acid, inhibit molds in liquid cosmetics and shampoos. **Calcium propionate** prevents mold growth in bread. All of these organic acids inhibit enzymatic or metabolic activity; their activity is not related to their acidity.

Aldehydes. Among the most effective antimicrobials are the **aldehydes** such as **formaldehyde**. In the form of an aqueous solution, this gas is called formalin and is used to preserve biological specimens and embalm bodies. Glutaraldehyde is a less irritating form; in a 2% solution (Cidex®), it is bacteriocidal, tuberculocidal, and virucidal in 10 minutes. It is sporicidal after about 10 hours of contact.

Gaseous chemosterilizers. **Ethylene oxide** inactivates protein by a mechanism similar to the aldehyes; that is, it combines with organic functional groups of proteins. It is a sterilant that kills all microbes and endospores and has great penetrating power. It is used to sterilize disposable medical supplies and equipment used in hospitals when heat treatment is not practical. **Propylene oxide** and **beta-propiolactone** are also important gaseous sterilants.

Oxidizing agents. **Hydrogen peroxide** is a common household antiseptic. It is not a good antiseptic for open wounds, but it is useful in deep anaerobic wounds where the oxygen released by the action of the enzyme catalase in tissue is effective against anaerobes such as *Clostridium*. **Zinc peroxide** and benzoyl peroxide are also useful in treatment of deep wounds, although the latter is better known as the main ingredient in many acne medications. **Ozone** is a highly reactive form of oxygen. It is generated by high-voltage electric discharges, and is a potential replacement for chlorination in water treatment.

Test

In the matching section, there is only one answer to each question; however, the lettered options (a, b, c, etc.) may be used more than once or, perhaps, not at all.

I. Match

1. A suffix meaning "to kill".

2. Destroying *all* forms of microbial life.

3. Absence of pathogens on an object or area.

4. Reduction of microbial populations to safe public health levels.

5. Chemical disinfection of living tissue such as skin or mucous membranes.

a. Disinfection
b. Sterilization
c. Antisepsis
d. Asepsis
e. Sanitization
f. Degerming
g. -cide
h. -stat

II. Match

1. Lowest temperature required to kill a liquid culture of a certain species of bacteria in 10 minutes at pH 7.

2. The time in minutes required to kill 90% of a bacterial population.

3. Mild heating to destroy particular spoilage organisms or disease organisms in milk or similar products.

4. Test for effectiveness of a chemical disinfectant.

5. Absence of water, resulting in condition of dryness.

a. Thermal death time

b. Decimal reduction time

c. Thermal death point

d. Phenol coefficient

e. Pasteurization

f. Desiccation

g. Incineration

63

III. Match

1. Ethylene oxide.   _i_   a. Phenolic
2. Sodium hypochlorite.   _b_   b. Halogen
3. Copper sulfate.   _d_   c. Alcohol
4. Mercurochrome.   _d_   d. Heavy metal
5. Benzalkonium chloride.   _e_   e. Quaternary ammonium compounds
6. Acid-anionic detergents. _f_   f. Surface-active agents
7. Sorbic acid.   _g_   g. Organic acids
8. Benzoyl peroxide.   _j_   h. Aldehydes
9. Hexachlorophene.   _a_   i. Gaseous chemosterilizers
10. Isopropanol.   _c_   j. Oxidizing agent(s)
11. Ozone.   _j_

IV. Fill-in-the-Blanks

1. Ultraviolet light is an example of _____nonionizing_____ radiation.

2. Sunlight owes its germicidal activity mainly to the formation of _____singlet_____ oxygen.

3. A good example of ionizing radiation is _____X ray (gamma ray)(high E e⁻)_____

4. A _____tincture_____ of iodine, for example, means an aqueous alcohol solution.

5. An _____Iodophor_____ is a combination of iodine and an organic molecule.

6. Chloramines consist of chlorine and _____ammonia_____.

7. Ethanol is usually used in a concentration of about _____70%_____.

8. Formaldehyde in an aqueous solution is called _____formalin_____.

9. _____Ozone_____ is a highly reactive form of oxygen generated by high-voltage electric discharges.

10. A less irritating form of formaldehyde is _____glutaraldehyde_____

11. A compound that would only inhibit the growth of a fungus would be a fungi-_____stat_____.

12. Steam _____heat____ under pressure_____ allows temperatures above boiling to be reached.

13. ___*Moist Autoclave*___ heat is generally more efficient.

14. Ultraviolet light causes bonds to form between adjacent thymines in ___*DNA*___.

15. Pseudomonads are often resistant to antimicrobials because of the small size of their cell wall ___*porins*___.

16. *Staphylococcus aureus*, *Pseudomonas aeruginosa*, and *Salmonella typhi* are tested under standard conditions for susceptibility to a test disinfectant and phenol in the ___*Phenol coeffic*___ test.

17. The most effective wavelength of ultraviolet light for germicidal activity is about ___*260*___ nm.

18. Generally speaking, the group of organisms that is more resistant to osmotic pressure than bacteria is ___*fungi (ak mold + yeast)*___.

19. Chemicals may denature proteins by reacting with their ___*disulfide (S-H)*___ bonds.

20. A liquid chemical disinfectant that is sporicidal yet requires more than 10 hours of contact is ___*Glutar aldehyde (Cidex)*___.

21. An antimicrobial found in many deodorant soaps that is mainly effective against gram-positive bacteria is ___*Triclo Carbon*___.

Key

Matching

I. 1.g 2.b 3.d 4.e 5.f
II. 1.c 2.b 3.e 4.d 5.f
III. 1.i 2.b 3.d 4.d 5.e 6.f 7.g 8.j 9.a 10.c 11.j

Fill-in-the-Blanks

1. nonionizing   2. singlet   3. x-rays, gamma rays, high energy electrons   4. tincture
5. iodophore   6. ammonia   7. 70%   8. formalin   9. ozone   10. glutaraldehyde
11. *stat*   12. under pressure (as in autoclave)   13. moist   14. DNA   15. porins
16. phenol coefficient   17. 260   18. fungi such as molds and yeasts
19. disulfide (-SH)   20. Glutaraldehyde (Cidex®)   21. Triclocarbon

8 *Microbial Genetics*

Chromosomes are cellular structures made up of genes that carry hereditary information. **Genetics** is the study of how genes carry information, are replicated and passed to other generations, and how they affect the characteristics of an organism.

Structure and Function of the Genetic Material

In Chapter 2 we saw that DNA is composed of repeating **nucleotides** containing the bases adenine (A), thymine (T), cytosine (C), or guanine (G); a deoxyribose sugar; and a phosphate group. Bases occur in specific complementary pairs, the hydrogen bonds from which connect strands of DNA: adenine with thymine, and cytosine with guanine. A **gene** is a segment of DNA that codes for a functional product. The information in DNA can be transcribed into RNA (transcription), and this information, in turn, translated into protein (translation).

Genotype Chromosomes

The **genotype** is an organism's genetic makeup, the information that codes for all the characteristics and potential properties of the organism. The genotype is its gene collection—its DNA. The **phenotype** refers to an organism's actual expressed properties, such as its ability to perform a chemical reaction. The phenotype is the collection of enzymatic or structural proteins.

DNA and Chromosomes

DNA in chromosomes is in the form of one long double helix. In procaryotes DNA is not found within a nuclear membrane. The eucaryotic chromosome is complexed with histone proteins, which are not found in procaryotes.

DNA Replication

In DNA replication (Figure 8-2 in the text), the two helical strands unravel and separate from each other at a **replication fork** (Figure 8-5 in the text), where the synthesis of new strands begins. The complementary pairing of bases—for example, adenine with thymine—yields a complementary copy of the original DNA. Segments of new nucleotides are joined to form short strands of DNA by **DNA polymerase** enzymes. Short strands of DNA are then joined into continuous DNA by action of **DNA ligase** enzymes. Since each new double-stranded DNA molecule has one original strand and one new strand, the process is called **semi-conservative replication.**

RNA and Protein Synthesis

Transcription

In **transcription**, a strand of **messenger RNA** (**mRNA**) is synthesized from the genetic information in DNA. (Adenine in the DNA dictates the location of uracil, which replaces thymine in mRNA). If DNA has the base sequence ATGCAT, the mRNA will have UACGUA. Only one DNA strand, the **sense strand**, serves as an RNA template. The region where RNA polymerase (needed for synthesis) binds to DNA and transcription begins is known as the **promoter site**. The terminator site is where the RNA polymerase and newly formed mRNA are released.

Translation

To form proteins from mRNA information (**translation**), one end of the mRNA becomes associated with a **ribosome**. Ribosomes consist of two subunits and contain protein and **ribosomal RNA** (**rRNA**). In the cytoplasm is a pool of 20 different amino acids. These are activated, in turn, by **transfer RNA** (**tRNA**); there is a different tRNA type for each amino acid type. The tRNA-amino acid attachment is made using an *amino acid activating enzyme* and energy from ATP. Also on each tRNA is an **anticodon**, a sequence of three bases (such as UAC) that matches a set of three bases (such as AUG) on mRNA called a **sense codon**. The tRNA-amino acid unit is brought into position where mRNA attaches to the ribosome. The ribosome moves along the mRNA strand, and amino acids are joined into a protein strand in which the amino acid sequence is dictated by the codon sequence in the mRNA. The amino acids are joined by **peptide bonds**, and tRNA is recycled. Several ribosomes may be attached at one time to a strand of mRNA. An mRNA strand with several attached ribosomes is called a **polyribosome**.

The Genetic Code

Because there are 64 possible codons and only 20 amino acids, most amino acids are signaled by several codons. This is called **degeneracy** of the code. There are 61 sense codons (each coding for an amino acid in the synthesized protein) and 3 nonsense codons (coding for termination of synthesis of a protein). The start of protein synthesis is signaled by the **initiator codon**.

Regulation of Gene Expression in Bacteria

Repression, Induction and Attenuation

Repression and Induction. An **inducer** is a substance (substrate) whose presence results in the formation of, or increase in the amount of, an enzyme. Such enzymes are called **inducible enzymes**; this genetically controlled response is termed **enzyme induction**. (Lactase production in response to lactose is an example.) Genetic regulation that decreases enzyme synthesis is **enzyme repression**. Repression occurs if cells are exposed to a particular end product of a metabolic pathway.

Operon Model

Protein synthesis in bacteria is controlled by a system called the **operon model**. As an example there are three enzymes involved in uptake and utilization of the sugar lactose in *E. coli*. The genes for these enzymes, **structural genes**, are close together on the bacterial chromosome. There is also an **operator site** next to these structural genes, and more remotely located, a gene that codes for a **repressor protein**. The operator and promoter sites plus the structural genes, are the **operon**.

When lactose is absent, the repressor protein prevents the operator from making lactose-utilizing enzymes. When lactose is present some diffuses into the cells and binds with the repressor protein. The operator then induces the structural genes to produce enzymes to utilize lactose, an example of an **inducible enzyme** (Figure 8-11 in the text). Many genes are not regulated in this manner, but are **constitutive**, and usually represent functions needed for major life processes. An example of constitutive enzymes are those for utilization of glucose. Cells prefer glucose to lactose. But, if the level of glucose is too low the cell responds with a series of events that cause formation of the lactose operon, enabling the cell to grow on lactose. This phenomenon is termed **catabolic repression** or the **glucose effect**.

Attenuation

It is wasteful of bacteria to produce enzymes they do not need. Normally, a cell growing in an amino-acid free medium will have enzymes to synthesize 20 amino acids. If an amino acid is added, the enzymes to synthesize it are made in smaller amounts. This control mechanism is called **attenuation**.

Mutation: Change in the Genetic Material

A **mutation** is a change in the base sequence of DNA.

Types of Mutation

The most common mutation is a **base substitution**, or **point mutation**, in which a single base in DNA is replaced with a different one. Such a substitution is likely to result in the incorporation of an incorrect amino acid in the synthesized protein, a result known as a **missense mutation**. Such an error may create a terminator codon, which stops protein synthesis before completion, resulting in a **nonsense mutation**. Deletion or addition of base pairs results in a **frameshift mutation**. In this mutation, there is a shift in the "translational reading frame," (the three-by-three grouping of nucleotides), and a long stretch of missense and an inactive protein product result. **Spontaneous mutations** occur without the known intervention of mutation-causing agents. Many chemicals and radiation bring about mutations; these are called **mutagens**.

Chemical Mutagens

Nitrous acid is a **base pair mutagen**. It causes adenine to pair with cytosine instead of thymine. Other mutagens are **base analogs**, which are structurally similar to bases and are incorporated into DNA by error. Examples are 2-aminopurine and 5-bromouracil, which are analogs of adenine and thymine. Some antiviral drugs are base analogs. Examples of **frameshift mutagens**, many of which are carcinogens, are benzypyrene (found in smoke and soot), aflatoxin (a mold toxin), and acridine dyes.

Radiation

Ionizing radiation, such as x-rays and gamma rays, damage DNA and are mutagens. Ultra-violet light (an example of **non-ionizing radiation**) is another mutagen that affects DNA. Certain enzymes can repair ultraviolet damage in a process stimulated by visible light.

Frequency of Mutation

The **mutation rate** is the probability of a gene mutating each time a cell divides. DNA replication is very faithful, and only about once in 10 billion base pair replications does an error occur. Mutagens increase the rate of such errors 10 to 1000 times.

Identifying Mutations

Mutants, which occur at low rates, can be identified more easily with bacteria because bacteria produce very large populations very quickly. **Positive** (**direct**) **selection** is illustrated by plating out bacteria on a medium containing penicillin. Survivors, which are resistant mutants, can be isolated. Nutritional mutants called **auxotrophs** are unable to synthesize a nutritional requirement, such as an amino acid—an activity the parent type is capable of doing. Colonies growing on a master plate containing a complete medium can be transferred by **replica plating**. This is an example of **negative** (**indirect**) **selection**. A sterile velvet pad is pressed onto the master plate, and the colonies are transferred simultaneously to a **minimal medium**, which lacks essential nutrients such as the required amino acid. An auxotrophic mutant will fail to appear on the minimal medium.

Identifying Chemical Carcinogens

The **Ames Test** (Figure 8-19 in the text) is based on the ability of a mutated cell to mutate again and to revert to its original form. An auxotroph of *Salmonella*, which has lost the ability to synthesize the amino acid histidine, is plated out on minimal medium without histidine. The test chemical (together with a rich source of activation enzymes found in rat liver extract) is placed on this plate also. Mutations of the *Salmonella* to the normal histidine-synthesizing form are indicated by colonies growing near the test chemical. High mutation rates are characteristic of carcinogens.

Genetic Transfer and Recombination in Procaryotes

Genetic recombination is the rearrangement of genes to form new combinations. If two chromosomes break and are rejoined in such a way that some of the genes are reshuffled between the two chromosomes, the process is called **crossing over**. The original chromosomes each have been recombined so that they carry a portion of genes from the other chromosome. Recombination is more likely to be beneficial than mutation and can occur in several ways. In all cases, however, the donor cell gives a portion of its total DNA to a different **recipient cell**. The **recipient**, called the **recombinant**, has DNA from the donor added to its own DNA.

Transformation in Bacteria

In **transformation**, "naked" DNA in solution is transferred from one bacterial cell to another. The process occurs naturally among very few genera of bacteria, and it usually occurs when donor and recipient are closely related and in the log phase of growth. To take up the DNA fragment, the recipient cell must be **competent**; that is, its cell wall must be permeable or possess specific receptor sites.

Conjugation in Bacteria

Conjugation requires contact between living cells of opposite mating types. The donor cell, called an F$^+$ cell (corresponding to maleness), has extra DNA pieces called **F factors**. These are a type of **plasmid** (discussed shortly), a free genetic element in the cell. When F+ and F- (corresponding to femaleness) cells are mixed, the F+ cells attach to sex pili to F- cells. F factors are duplicated by the donor, and the new copy is transferred to the F- cell, which becomes an F+ cell. The bacterial chromosome of the F+ is not passed, and no recombinants are produced. However, some F+ cells have the F factor integrated into their chromosome and are called **Hfr (high frequency of recombination) cells**. The chromosome of these cells may be transferred along with the integrated F factor if conjugation proceeds without interruption (a copy is retained by the donor). In this case the recipient cell becomes an Hfr donor cell. Considerable amounts of the donor chromosome can be transferred even if interruption occurs. In this case the recipient cell may acquire a sizeable portion of the donor's genome, but unless conjugation is completed, the F factor will not be transferred.

Transduction in Bacteria

In **generalized transduction**, the **phage** (short for "bacteriophage," a bacterial virus) attaches to the bacterial cell wall and injects DNA into the bacterium. Normally this directs the synthesis of new viruses. Sometimes, however, bits of bacterial chromosome are accidentally incorporated into the viral DNA. When these viruses infect a new host cell, they can incorporate this genetic information from the previous bacterial host into the new bacterial host. In **specialized (restricted) transduction** only certain bacterial genes are transferred.

Recombination in Eucaryotes

Eucaryotes differ from procaryotes in that genetic recombination is the result of a sexual reproductive process. This process involves the fusion of the haploid nuclei of two parental cells. The **haploid cells** (containing only one of each type of chromosome) are called **gametes** (male gametes are sperm; female gametes are ova). These produce a **zygote**, with half of the chromosomes from one parent and half from the other. Before reproducing sexually, the eucaryote must reduce its chromosome numbers from diploid to haploid by **meiosis**. Crossing over (Figure 8-20 in the text), by which portions of the chromosome may be exchanged, may occur during this stage.

Genetic Engineering

The manipulation of genetic material in the laboratory is called genetic engineering. **Plasmids**, circular pieces of DNA that replicate independently from the cell's chromosome, usually carry only genes not essential for growth of the cell. **Conjugative plasmids** (such as Hfr) may have genes for carrying out their transfer to another cell, and even the bacterial chromosome and other plasmids. **Dissimilation plasmids** code for enzymes to utilize unusual sugars and hydrocarbons. **Bacteriocinogenic plasmids** cause the synthesis of bacteriocins, toxic proteins that kill other bacteria. **Resistance factors (R-factors)** carry genes for antibiotic resistance and other antimicrobial factors. Sometimes R-factors contain two groups of genes; **resistance transfer factor**, coding for plasmid replication, and **r-determinant**, coding for enzymes that inactivate antimicrobials.

Transposons

Transposons (transposable genetic elements) are small segments of DNA that can move from one region of the chromosome to another (jumping genes). Such movement is uncommon, about the same as spontaneous mutation rates in bacteria. The simplest transposons are called **insertion sequences**; they code only for an enzyme (transposase) that cuts and ligates DNA in recognition sites. **Recognition sites** are short regions of DNA that the enzyme recognizes as recombination sites between transposon and chromosome. **Complex transposons** may carry other genes, such as those conferring antibiotic resistance. Transposons are a powerful mechanism for movement of genes from one chromosome to another, even between species.

Recombinant DNA

Recombinant DNA results from joining DNA from different sources, even different species, in vitro. Introduced into a bacterium, the bacterium can often synthesize the products of the new recombinant gene. Recombinant DNA is made with **restriction enzymes** that cut up DNA into short sequences. Such pieces have single-stranded self-complementary ends called sticky ends. Sticky ends can join the short pieces of DNA made by a given restriction enzyme or join with each other with DNA ligase. (Figure 8-29 in the text illustrates construction of a recombinant plasmid with this technique.) These plasmids can be introduced into a bacterium and induce it to produce useful products that it would otherwise be unable to synthesize.

Eucaryotic genes are made up of **exons** that code for their product and **introns** that are intervening regions that do not code for protein. Because procaryotic bacterial cells do not contain enzymes to deal with intron RNA, as do eucaryotic cells, they cannot make protein from such a eucaryotic gene. It is necessary instead to make a **complementary DNA (cDNA)** that contains only the protein coding exons of interest, but not the introns. By this means the bacterial cell can be made to transcribe and translate eucaryotic cDNA.

Test

In the matching section, there is only one answer to each question; however, the lettered options (a, b, c, etc.) may be used more than once or, perhaps, not at all.

I. Match

1. Where the RNA polymerase and the newly formed mRNA are released.

2. Enzymes that assemble the nucleotides of DNA into chains.

3. Formation of protein from the genetic information contained in mRNA.

4. Formation of mRNA from genetic information in DNA.

5. Enzymes that bind short strands of DNA together into longer strands.

6. Where transcription begins on mRNA.

7. Enzymes that chop plasmid-DNA into short fragments useful for genetic engineering.

a. DNA ligases

b. DNA polymerases

c. Transcription

d. Translation

e. Promoter site

f. Terminator site

g. Restriction enzymes

II. Match

1. A sequence of three bases coding for the position of an amino acid in the assembly of a protein chain.

2. A cluster of related genes together with the operator and promoter sites on mRNA.

3. A sequence of three bases on tRNA that locates the codon location on the mRNA at the ribosome.

4. A sequence of bases that does not code for an amino acid, but that terminates the protein chain.

a. Anticodon

b. Nonsense codon

c. Codon

d. Operon

e. Initiator codon

III. Match

1. The actual template upon which the protein chain is assembled.

2. The product of transcription.

3. One of these is specific for each of the 20 amino acids.

4. The original genetic information in a bacterial cell.

a. mRNA
b. tRNA
c. rRNA
d. DNA
e. cDNA

IV. Match

1. Probability of a gene mutation each time a cell divides.

2. Usually a result of the deletion or the addition of a base pair.

3. Mutation caused by a chemical that is structurally similar to nucleotide components such as adenine or thymine.

4. A mutagen that would, for example, make the base adenine pair with cytosine instead of thymine.

a. Missense mutation
b. Frameshift mutation
c. Spontaneous mutation
d. Base-pair type of mutagen
e. Base-analog type of mutagen
f. Base-substitution type of mutagen
g. Mutation rate

V. Match

1. DNA transferred between cells is in solution in the suspending medium.

2. Requires contact between living cells of opposite mating types.

3. Requires a sex pilus.

4. Hfr cells.

5. Method by which plasmids such as F factors are transferred between cells.

a. Conjugation in bacteria

b. Transformation in bacteria

c. Transduction in bacteria

VI. Match

1. Contain genes coding for enzymes that catabolize unusual sugars or hydrocarbons, for example.

2. Contain genes for synthesis of toxic proteins lethal for other bacteria.

a. Dissimilation plasmids

b. Bacteriocinogenic plasmids

c. Conjugative plasmids

VII. Fill-in-the-Blanks

1. An example of nonionizing radiation is _____.

2. A nutritional mutant is known as a _____.

3. Colonies growing on a master plate containing a complete medium can be transferred simultaneously to minimal medium by the _____ technique.

4. When accumulations of end products result in enzyme repression, the end product is the _____.

5. The *Salmonella* organism in the Ames test has lost the ability to synthesize the amino acid _____.

6. The site at which the replicating DNA strands separate is called the _____.

7. Several ribosomes may be attached at one time to an mRNA strand; this structure is called a _____.

8. The fact that there is more than one possible codon for each amino acid is called _____ of the code.

9. The organism's entire genetic potential is the _____.

10. A cell with a cell wall permeable to soluble DNA that has specific receptor sites for it is _____.

11. Bacteria that have the F factor integrated into their chromosome and that tend to transfer F factor and chromosome together are called _____ cells.

12. The haploid cells in a eucaryote are called the _____.

13. Reduction of chromosome numbers before a eucaryote reproduces sexually is known as _____.

14. Enzymes that are always present in the cytoplasm are called _____ enzymes.

15. In the operon model, the regulator gene codes for a _____ protein.

16. The place on the mRNA at which the repressor binds to prevent transcription of structural genes into protein is known as the _____ site.

17. Some R factors have a set of genes that codes for resistance, and another, the _____ that codes for replication and conjugation.

18. The regions on the chromosome of a eucaryotic cell that do not code for any protein are called _____.

19. A bacterial virus is known, for short, as a _____.

20. In making a new strand of DNA, where there is an adenine on the original strand there will be _____ on the new strand.

21. In making a strand of mRNA from DNA, where there is an adenine on the DNA there will be _____ on the RNA.

22. Small segments of DNA that can move from one region of the chromosome to another are called _____.

23. Recognition sites are short regions of DNA that transposase recognizes as recombination sites between _____ and chromosome.

Key

Matching

I.   1.f 2.b 3.d 4.c 5.a 6.e 7.g
II. 1.c 2.d 3.a 4.b
III. 1.a 2.a 3.b 4.d
IV. 1.g 2.b 3.e 4.d
V.   1.b 2.a 3.a 4.a 5.a
VI. 1.a 2.b

Fill-in-the-Blanks

1. ultraviolet light   2. auxotroph   3. replica plating   4. corepressor   5. histidine
6. replication fork   7. polyribosome   8. degeneracy   9. genotype   10. competent
11. Hfr   12. gamete   13. meiosis   14. constitutive   15. repressor   16. operator
17. resistance transfer factor   18. intron   19. phage   20. thymine   21. uracil
22. transposons   23. transposon

9 Classification of Microorganisms

The science of classification, especially of living forms, is called **taxonomy**.

Five-Kingdom System of Classification

In 1969, H.R. Whittaker proposed the **five-kingdom system** of biological classification. The main division in the system is made between procaryotic and eucaryotic cells. Included with the procaryotic organisms (called **Monera** and **Procaryotae**) are the bacteria and cyanobacteria (formerly called blue-green algae). In the classification of eucaryotic cells, the simpler organisms, mostly unicellular, are grouped as **Protista** and include funguslike water molds, slime molds, protozoans, and the more primitive algae. Branching from the Protista are three kingdoms of more complex organisms distinguished primarily by their nutritional mode. **Fungi** are eucaryotic organisms that include yeasts, molds, and mushrooms. These absorb dissolved organic matter. **Plantae** are multicellular algae, mosses, ferns, conifers, and flowering plants. These obtain energy by photosynthesis. **Animalia** are multicellular animals and include sponges, various worms, insects, and vertebrate animals. These obtain nutrition by ingestion of organic matter, as by mouth.

A Three-Kingdom System

In 1978 Carl Woese proposed a three-kingdom system which is based upon modern techniques in molecular biology and biochemistry. Primarily, the distinction is that there are two types of procaryotic cells, the **eubacteria** and the **archaeobacteria**. The third kingdom would be all of the eucaryotes.

Archaeobacteria differ from eubacteria in that their cell walls never contain peptidoglycan, they live in extreme environments and carry out unusual metabolic processes: the anaerobic methanogens, extreme halophiles requiring high concentrations of salt, and thermoacidophiles found in hot, acidic environments.

Classification of Organisms

Scientific Nomenclature

The system of scientific nomenclature used today was developed in the eighteenth century by Carolus Linnaeus. All organisms have two names; the **genus name** and the **specific epithet**. Both names appear underlined or italicized, and the genus name is always capitalized. This system of two names for an organism is called **binomial nomenclature**.

The Taxonomic Hierarchy

For eucaryotic organisms, a **species** is a group of closely related organisms that breed among themselves and not with other species. A **genus** (plural is *genera*) consists different species that are related by descent. Related genera make up a **family**. A group of similar families constitutes an order, and a group of similar orders makes up a **class**. Related classes in turn comprise a **division**. (In zoology **phylum** is a term comparable to division.) All phyla or divisions that are related to each other make up a kingdom. Organisms are arranged into taxonomic categories called taxa, which reflect degrees of relatedness among organisms. The hierarchy of taxa shows evolutionary or **phylogenetic** (from a common ancestor) relationships.

Classification of Bacteria

The term *species*, as applied to higher organisms, does not apply well to bacteria. Bacterial morphological traits are limited and sexuality involves unilateral transfer of genetic material rather than breeding, which is the fusion of egg and sperm. A **bacterial species** is, accordingly, defined as a population of cells with similar characteristics. Sometimes a pure culture of the same species contains distinguishable groups within the species that are called **strains**. Presumably, these are derived from a single cell and are identified by numbers, letters, or names following the species name.

Bergey's Manual of Systematic Bacteriology is the standard reference for bacterial taxonomic classification. It is being issued in four volumes, in which bacteria are divided into four groups. One of these is the archaeobacteria. Until all volumes are complete, *Bergey's Manual of Determinative Bacteriology*, 8th Edition can be used. In *Bergey's Manual*, bacteria are grouped according to shared properties such as morphology, staining characteristics, nutrition, and metabolism.

Classification of Viruses

Viruses (see Chapter 12) are not classified under any of the kingdoms by which bacteria are grouped.

Classification and Identification of Microorganisms

Morphological Characteristics

Literally hundreds of bacterial species appear in the shape of small rods or small cocci. Nonetheless, these characteristics are still of some use in classifying bacteria. Sometimes the presence of endospores or flagella can be helpful. One of the first steps in identification is doing a Gram stain. Other stains, such as the acid-fast stains, are sometimes useful for several bacterial groups.

Biochemical Tests

Enzymatic activity is widely used to differentiate among bacteria. Tests that determine the ability to ferment an assortment of selected carbohydrates are particularly common. Bacteria also may be subjected to other biochemical tests. The use of selective or differential media (Chapter 6) has been discussed for use in bacterial isolation or identification.

Serology

Microorganisms are antigenic; that is, they are capable of stimulating antibody formation when injected into animals. The immune system of an animal injected with a bacterial species will produce antibodies highly specific for that species of bacteria. Solutions of such antibodies, called **antisera** (singular is **antiserum**), are commercially available for identification of many medically important bacteria. In a demonstration of such methods, an unknown bacterium is placed in a drop of salt solution on a microscope slide and mixed with a drop of known antiserum. In such a **slide agglutination test**, the bacteria will agglutinate, or clump together, when mixed with antibodies produced against that strain or species.

Phage Typing

Phages (Chapter 8), like antibodies, are highly specific for species, or strains within a species. By means of stock of different phages, this characteristic can be used to identify bacteria or even to trace epidemics. **Phage typing** is commonly used in outbreaks of illness causd by staphylococci.

Amino Acid Sequencing

The amino acid sequence of a protein is a direct reflection of the base sequence of the encoding gene. Thus, by comparing the amino acid sequence of proteins from two different organisms, it is possible to determine the degree of their relatedness. The more similar the proteins, the more closely related are the organisms.

Protein Analyses

Protein profiles are obtained by using **polyacrylamide gel electrophoresis (PAGE)**. Disintegrated cell proteins are distributed by an electric current (electrophoresis) in a gel. The rate of movement, which is visible by staining, "fingerprints" the cells' proteins and can show relationships between bacterial species.

Base Composition of Nucleic Acids

A technique that may be able to suggest evolutionary relationships among bacteria is the determination of the base composition of the DNA, the **G + C ratio**. The number of guanine (G) and cytosine (C) base pairs can be determined as a percentage. (By inference, this will give the percentage of adenine (A) and thymine (T) as well. If there is a difference of more than 10% in the percentage of G + C pairs, the organisms are probably unrelated. On the other hand, two organisms with the same percentage of G + C need not be closely related; other supporting data are needed.

Nucleic Acid Hybridization

If a double-stranded molecule of DNA is heated, the complementary strands will separate as the hydrogen bonds connecting the bases break. However, because the strands are complementary, they will reunite if allowed to incubate together. If the separated strands of two different organisms are incubated together, the closer the relationship, the greater the amount of DNA that will join together by complementary base pairing. This technique is known as **nucleic acid hybridization** and assumes that if two species are similar or related, a major portion of their nucleic acid sequences also will be similar. **DNA probes**, based on these techniques, are being developed for rapid identification of bacteria.

Genetic Recombination

Figure 9-15 in the text illustrates a technique for identifying an unknown bacterium. Extracted DNA from the unknown species is incubated with an auxotrophic mutant and then cultured on minimal media designed to suppress the growth of the auxotrophic mutant. If the two species are the same or closely related, then some transformation will have occurred and growth will be observed. No growth indicates that the bacteria are probably not related.

Numerical Taxonomy

In **numerical taxonomy**, which is used to help discover relationships among organisms, many microbial characteristics are listed, and the presence or absence of each characteristic is scored for each organism. (Weakness of the method is that all characteristics are given equal weight.) When different bacteria are matched against each other (a computer is helpful), the greater the number of characteristics shared, the closer the taxonomic relationship is assumed to be. A match of 90% or more usually indicates a single taxonomic unit or species.

Test

In the matching section, there is only one answer to each question; however, the lettered options (a, b, c, etc.) may be used more than once or, perhaps, not at all.

I. Match

1. Bacteria.

2. Elephants.

3. Multicellular algae.

4. Mushrooms.

5. Yeasts.

6. Cyanobacteria.

a. Monera

b. Protista

c. Animalia

d. Plantae

e. Fungi

II. Match

1. A distinguishable group within a species of bacteria.

2. Made up of all phyla or divisions that are related to. each other

3. In bacteria, all cells with similar characteristics. each other.

4. In eucaryotes, closely related organisms that breed among themselves.

5. A group of similar orders in eucaryotic cells.

a. Strain

b. Species

c. Genus

d. Kingdom

e. Class

f. .Order

III. Match

1. A serological test for bacterial identification.

2. Makes use of bacterial viruses to identify bacteria.

3. Accumulation of a microbe's characteristics, each with equal weight. The more characteristics shared, the closer the taxonomic relationship.

4. Determines the taxonomic relationship of bacteria by allowing complementary strands of DNA from different organisms to reassemble as a complementary pair.

5. Taxonomic relationship of bacteria found by base pairs percentage in DNA.

6. Used in analysis of bacterial protein.

a. Numerical taxonomy

b. Phage typing

c. Nucleic acid hybridization

d. Amino acid sequencing

e. G + C ratio

f. Slide agglutination test

g. Polyacrylamide gel electrophoresis

IV. Fill-in-the-Blanks

1. The science of classification, especially of living forms, is called _____.

2. The system we use for naming biological organisms with two names is called
   _____.

3. The taxonomic categories into which organisms are arranged and that reflect degrees of relatedness among them are called _____.

4. The taxonomic classification scheme for bacteria may be found in the book called
   _____ *Manual.*

5. _____ is a technique commonly used to trace the epidemiology of outbreaks of illness caused by staphylococci.

6. The heirarchy of taxa shows evolutionary or _____ (from a common ancestor) relationships.

7. Solutions of antibodies against specific bacteria are called _____.

8. A technique that may be able to suggest evolutionary relationships among bacteria is the determination of the base composition of the DNA, the _____ ratio.

9. The plural of genus is _____.

10. The cell walls of archaeobacteria differ from those of eubacteria by a lack of
    _____.

Key

Matching

I.   1.a 2.c 3.d 4.e 5.e 6.a
II. 1.a 2.d 3.b 4.b 5.e
III. 1.f 2.b 3.a 4.c 5.e 6.g

Fill-in-the-Blanks

1. taxonomy   2. binomial nomenclature   3. taxa   4. *Bergey's*   5. phage typing
6. phylogenetic   7. antisera   8. G + C ratio   9. genera   10. peptidoglycan

10 *Bacteria*

Bacterial Groups

The Spirochetes

Spirochetes have a helical shape and are motile by means of **axial filaments**. One end of each axial filament is attached near a pole of the cell (Figure 10-1). Axial filaments are wound around the body of the cell, and the cell moves by stretching and relaxing the axial filament. *Treponema pallidum*, the cause of syphilis; the genus *Borrelia*, which causes relapsing fever; and *Leptospira* species, which cause leptospirosis, are all spirochetes.

Aerobic/Microaerophilic, Motile Helical/Vibrioid, Gram-Negative Bacteria

Some helical bacteria are not included with the spirochetes because they lack an axial filament. They are motile instead by means of flagella, which may occur singly or in tufts at either or both poles of the cell. Many of these bacteria are harmless aquatic organisms such as *Spirillum volutans*, which are probably adapted to the low concentrations of organic matter in their habitat. *Azospirillum* bacteria fix nitrogen in close assocation with the roots of grasses. Among pathogens in this group, *Campylobacter fetus* causes abortion in domestic animals, and *Campylobacter jejuni* causes foodborne enteritis. **Vibrioid** is the term applied to helical bacteria that do not have a complete turn, such as *Bdellovibrio* spp., which attacks other bacteria.

Gram-Negative, Aerobic Rods and Cocci

The genus *Pseudomonas* is characterized by a polar flagellum. Many species excrete extra-cellular, water soluble pigments. *Pseudomonas aeruginosa* produces a blue-green pigment and can infect the urinary tract, burns, and wounds. Pseudomonads are common in soil, and can often grow at refrigerator temperatures. They are able to decompose chemicals such as pesticides in soil, and their ability to grow in antiseptic solutions and whirlpool baths as well as their resistance to antibiotics make them troublesome. Many pseudomonads substitute nitrate for oxygen during anaerobic respiration, and so deplete nitrate fertilizers of nitrogen, which escapes in the form of nitrogen gas.

Legionella is the cause of legionellosis. *Azotobacter* and *Azomonas* are nitrogen-fixing soil organisms. *Rhizobium* and *Bradyrhizobium* form a nitrogen-fixing symbiotic relationship with

Figure 10-1. Spirochetes are helically shaped bacteria that are motile by means of an axial filament. There are at least two axial filaments per cell. Each is anchored to one end of the cell but not to the other. Causing the cell to rotate in a corkscrew fashion and flexing movement of the axial filaments can cause locomotion of the cell in fluids.

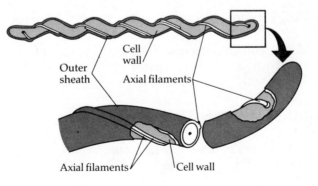

legumes. *Acetobacter* and *Gluconobacter* convert ethanol into vinegar. *Zoogloea* forms flocculant slimy masses essential to operation of activated-sludge sewage treatment systems. *Neisseria gonorrhea* causes gonorrhea, and *Neisseria meningitidis* causes meningococcal meningitis. *Moraxella lucanata* causes conjunctivitis. *Brucella* causes brucellosis, and has the unusual ability to survive phagocytosis. *Bordetella pertussis* causes whooping cough. *Francisella tularensis* causes tularemia.

Facultatively Anaerobic, Gram-Negative Rods

There are three families in this group, which we shall discuss in turn.

Family Enterobacteriaceae (enterics). The **enterics** are clinically important, and there are many biochemical tests used (such as the IMViC tests) to distinguish between them. Many enterics have pili for attachment to mucous membranes and sex pili for conjugation. Many also produce proteins called **bacteriocins**, which kill related species of bacteria.

Escherichia coli is one of the most common inhabitants of the intestinal tract and a familiar laboratory bacterium. It is used as an indicator organism for fecal pollution. It can cause urinary tract infections. Occasionally enterotoxins produced by *E. coli* cause traveler's diarrhea.

Almost all members of the genus *Salmonella* are pathogenic. Typhoid fever is caused by *Salmonella typhi*. Taxonomically the salmonellae are divided into hundreds of **serovars** (serotypes) by serological means; no individual species are recognized in spite of their species-like name. Many are known by antigenic formulas, according to the Kauffmann-White scheme. For example, *Salmonella typhimurium* is represented as 1, 4 (5), 12, i, 1, 2.

Species of *Shigella* cause bacillary dysentery or shigellosis; *Klebsiella pneumoniae* causes a form of pneumonia; and *Serratia marcescens* produces a red pigment and causes infections of the urinary and respiratory tracts in hospital patients. Members of the genus *Proteus* are actively motile and may cause infections of urinary tracts and wounds, as well as infant diarrhea. *Yersinia pestis* causes bubonic plague. *Erwinia* species are primarily plant pathogens, producing enzymes that dissolve the pectin between individual plant cells (soft-rots). *Enterobacter cloacae* and *Enterobacter aerogenes* cause urinary tract infections. Both are **non-fecal coliforms**.

Family Vibrionaceae. Morphologically, bacteria in the genus *Vibrio* are characterized as slightly curved rods. Important pathogens are *Vibrio cholerae*, the cause of cholera, and *Vibrio parahaemolyticus*, the cause of a form of gastroenteritis transmitted mostly by shellfish.

Family Pasteurellaceae. The organisms of the genus Pasteurella are mostly pathogens of domestic animals. *Pasteurella multocida* can be transmitted to humans by dog and cat bites. *Hemophilus influenzae* is a common cause of meningitis, otitis media, epiglottitis, and a number of other important diseases. *Hemophilus* bacteria are cultured on media enriched by hemoglobin containing X and V factors. *Gardnerella vaginalis* causes vaginitis.

Anaerobic, Gram-Negative, Straight, Curved, and Helical Rods

Members of the genus *Bacteroides* live in the human intestinal tract. Some are found in the oral cavity, genital tract, and upper respiratory tract. They do not form endospores. *Fusobacterium* are long and slender with pointed ends. In humans they often cause dental abscesses.

Dissimilatory Sulfate-Reducing or Sulfur-Reducing Bacteria

These organisms are not medically important. They are obligately anaerobic bacteria that use oxidized forms of sulfur as electron acceptors, which produces hydrogen sulfide. They are found in anaerobic sediments and human and animal intestinal tracts. *Desulfovibrio* is the best known genus.

Anaerobic, Gram-Negative Cocci

Bacteria of the genus *Veillonella* are part of the normal mouth flora and are components of dental plaque.

The Rickettsias and Chlamydias

Both of these are obligate, intracellular parasites, like viruses, but in morphological and biochemical aspects they resemble, and are classified as, bacteria. They are usually pathogenic.

Rickettsias are gram-negative. They are often transmitted to humans by insects or ticks. *Coxiella burnetii* causes Q fever. Other rickettsial diseases are epidemic typhus caused by *Rickettsia prowazekii*, endemic murine typhus caused by *Rickettsia typhi*, and Rocky Mountain spotted fever caused by *Rickettsia rickettsii*. They are usually cultivated in the yolk sac of chicken embryos.

Chlamydia do not require arthropods (such as insects) for transmission. The infectious form, the **elementary body**, attaches to a host cell and is phagocytized and housed in a cell vacuole. Within the host cell the elementary body becomes a larger, less infective reticulate body that divides successively. Eventually it condenses into infectious elementary bodies that are released to infect surrounding host cells. *Chlamydia trachomatis* causes trachoma, as well as the venereal diseases non-gonococcal urethritis and lymphogranuloma venereum. *Chlamydia psittaci* causes psittacosis (ornithosis). Chlamydia are cultivated laboratory animals, cell cultures, or in the yolk sac of chicken embryos.

The Mycoplasmas

Mycoplasmas are bacteria that do not form cell walls. They cannot revert to normal cell-wall-containing bacteria as can L forms. *Mycoplasma pneumoniae* causes primary atypical pneumonia. *Spiroplasma* are serious plant pathogens. *Ureaplasma* may cause urinary tract infections. *Thermoplasma*, which are also grouped with archaeobacteria, are sometimes found in hot water systems. To grow mycoplasmas on artificial media, sterols must be provided.

Gram-Positive Cocci

Staphylococci occur in grapelike clusters. The most important species is *Staphylococcus aureus*. Staphylococci are able to survive and grow at high osmotic pressures. *S. aureus* produces a yellow pigment and many toxins. Among the toxins is an enterotoxin that causes food poisoning. Rapid development of antibiotic resistance is another troublesome characteristic.

Streptococci cause a great variety of diseases. They do not use oxygen, but are mostly aerotolerant. **Alpha-hemolytic** species form a greenish zone around a colony on blood agar. **Beta-hemolytic** species form a clear zone of hemolysis. **Non-hemolytic** (gamma-hemolytic) streptococci are not hemolytic. The streptococci produce substances that destroy phagocytic cells, enzymes that digest the host's connective tissue and enzymes that lyse fibrin.

Endospore-Forming, Gram-Positive Rods and Cocci

Bacillus species are aerobic or facultative anaerobes. *Bacillus anthracis* causes anthrax. *Clostridium* species are mostly obligate anaerobes *Clostridium tetani* causes tetanus, *Clostridium botulinum* causes botulism, and *Clostridium perfringens* causes gas gangrene.

85

Regular, Non-Sporing, Gram-Positive Rods

The genus *Lactobacillus* represents a group of aerotolerant bacteria that produce lactic acid from simple carbohydrates and grow in acidic environments. They are found in the human vagina and oral cavity. Industrially, they produce products such as sauerkraut and yogurt. The pathogen *Listeria monocytogenes* survives within phagocytic cells, can grow at refrigerator temperatures, and can cause serious damage to a fetus.

Gliding, Sheathed, and Budding and/or Appendaged Bacteria

Appendaged Bacteria. This group of bacteria is linked taxonomically by the presence of prosthecae, which are protrusions such as stalks and buds. *Caulobacter* are the best studied, having stalks that anchor them to surfaces. This increases their nutrient uptake in the low-nutrient environments where they are found. *Caulobacter* reproduction results in one stalked cell and one flagellated swarmer cell, which eventually becomes another stalked cell.

Gliding, Nonfruiting Bacteria. Gliding bacteria are motile by gliding over surfaces. *Cytophaga* are important cellulose-degraders in the soil. *Beggiatoa* morphologically resembles cyanobacteria but is not photosynthetic. It uses hydrogen sulfide as an energy source.

Gliding, Fruiting Bacteria. The **myxobacteria** have a remarkable life cycle. Large numbers of vegetative cells converge upon a single point, where they aggregate and differentiate into a stalked body that carries resting cells called myxospores. These myxospores eventually germinate and renew the cycle.

Budding Bacteria. Budding bacteria form buds; that is, the parent cell retains its identity and the bud increases in size until it separates as a new cell. The best known are members of the genus *Hyphomicrobium*.

Sheathed Bacteria. Sheathed bacteria, such as *Sphaerotilus natans*, form a filamentous sheath that surrounds the bacteria. They are found in fresh water and sewage.

Irregular, Non-Sporing, Gram-Positive Rods

Corynebacterium diphtheriae causes diphtheria. (Coryne means "club," which reflects the irregular morphology of the genus.) *Propionibacterium acnes* is an anaerobic skin organism responsible for acne. *Actinomyces* spp. are filamentous, anaerobic bacteria that can fragment into coryneform cells. *Actinomyces israelii* causes actinomycosis.

Mycobacteria

The name myco suggests "fungus," so named because of the occasional filamentous growth of this rod-shaped genus. Most species are acid-fast. *Mycobacterium tuberculosis*, the cause of tuberculosis, and *Mycobacterium leprae*, the cause of leprosy, are important pathogens in this genus. A number of non-pathogenic soil organisms also are found among the mycobacteria.

Nocardioforms

Members of the genus *Nocardia* are filamentous bacteria that resemble the Actinomyces morphologically, but are aerobic. They are common in soil. *Nocardia asteroides* is a cause of pulmonary nocardiosis, a chronic lung infection, and mycetoma, a localized destructive infection of the feet or hands.

Chemoautotrophic Bacteria

These organisms are capable of using inorganic chemicals as energy sources and carbon dioxide as the only source of carbon. *Nitrobacter* oxidizes ammonium (NH_4^+) to nitrite (NO_2^-). *Nitrosomonas* oxidizes nitrite to nitrate (NO_3^-). Nitrate is mobile and the form of nitrogen best available to plants. For their source of energy *Thiobacillus* and other sulfur-oxidizing bacteria oxidize reduced forms of sulfur, such as hydrogen sulfide (H_2S) or elemental sulfur (S), into sulfates (SO_4^{2-}).

Archaeobacteria

This interesting group of organisms includes the extreme **halophiles** such as *Halobacterium* or *Halococcus* that actually require high concentrations of sodium chloride to grow. *Sulfolobus* also thrives in extreme environments such as acidic, sulfur rich, hot springs. The **methane-producing** bacteria that derive energy by combining hydrogen (H_2) with CO_2 to form methane (CH_4) are important in anaerobic sewage treatment.

Phototrophic Bacteria

There are three groups of bacteria that are **phototrophic**; that is, they use light as an energy source.

The **purple** or **green sulfur bacteria** are generally anaerobes that inhabit deep aquatic sediments. They are **anoxygenic**: their photosynthesis does not produce oxygen. Instead of splitting water (H_2O) to release oxygen, they split hydrogen sulfide (H_2S) to form elemental sulfur (S).

Other phototrophs, the **purple nonsulfur** and **green nonsulfur bacteria**, use organic compounds such as acids and carbohydrates for the photosynthetic reduction of carbon dioxide rather than using the hydrogen in water or in hydrogen sulfide.

The photosynthetic **cyanobacteria** are aerobes that carry out oxygen-producing photo-synthesis much like higher plants. Many species of cyanobacteria are capable of fixing nitrogen from the atmosphere. They do so using enzymes carried in structures called **heterocysts**. **Gas vacuoles** in many aquatic species help maintain buoyancy for best access to light. There are unicellular, colonial, and filamentous forms of cyanobacteria.

Actinomycetes

Actinomycetes have a mold-like growth habit with long, branched filaments. The diameter of the filaments is of bacterial dimensions. Most reproduce by forming asexual spores. One genus, *Frankia*, causes formation of nitrogen-fixing nodules on alder tree roots. *Streptomyces* form conidiospores at the ends of filaments. Strict aerobes, these bacteria are important degraders of proteins, starch, and cellulose in soil. They produce **geosmin**, a gas that gives soil its odor. Most commercial antibiotics are produced by *Streptomyces*.

Test

In the matching section, there is only one answer to each question; however, the lettered options (a, b, c, etc.) may be used more than once or, perhaps, not at all.

I. Match

1. Cause of syphilis.

2. Do not form cell walls.

3. An important genus in this group is *Pseudomonas*.

4. *Staphylococcus aureus*.

5. *Streptomyces* spp.

6. Motile by means of axial filament.

a. Spirochetes

b. Gram-negative aerobic rods and cocci

c. Gram-negative anaerobic bacteria

d. Gram-positive cocci

e. Actinomycetes

f. Mycoplasma

II. Match

1. Obligate intracellular parasites.

2. *Clostridium* spp.

3. *Bacteroides* spp.

4. Cause of gonorrhea.

5. *Lactobacillus* spp.

6. *Veillonella* spp.

7. *Escherichia coli*.

a. Helical/vibrioid, gram-negative bacteria
b. Facultatively anaerobic gram-negative rods
c. Gram-negative, aerobic rods and cocci
d. Anaerobic, gram-negative, straight, curved, and helical rods
e. Anaerobic, gram-negative cocci
f. Endospore-forming, gram-positive rods and cocci
g. Regular, non-sporing, gram-positive rods
h. Rickettsias and chlamydias

III. Match

1. *Vibrio* spp.

2. *Neisseria* spp.

3. *Streptococcus* spp.

4. *Streptomyces* spp.

5. *Chlamydia* spp.

6. *Hemophilus* spp.

7. *Gluconobacter* spp.

a. X and V factors

b. Cholera

c. Group A beta-hemolytic

d. Botulism, tetanus, gas gangrene

e. Form conidiospores on filaments

f. Form infectious elementary bodies

g. Gonorrhea

h. Ethanol into vinegar

IV. Match

1. Cause of whooping cough.

2. Produces a food poisoning enterotoxin.

3. Rickettsial organism.

4. Bubonic plague.

5. Water-soluble, blue-green pigment.

6. Essential for operation of activated-sludge sewage system.

7. Gliding bacterium.

8. A filamentous bacterial pathogen.

a. *Pseudomonas aeruginosa*

b. *Bordetella pertussis*

c. *Escherichia coli*

d. *Yersinia pestis*

e. *Staphylococcccus aureus*

f. *Coxiella burnetti*

g. *Zoogloea* spp.

h. *Nocardia asteroides*

i. *Cytophaga* spp.

V. Match

1. A genus of gliding bacteria that is an important cellulose degrader.

2. A sheathed bacterium found in water and sewage.

3. A chemoautotrophic bacterium that participates in nitrification in soil.

4. Photosynthetic bacteria that may fix nitrogen.

5. Photosynthetic, anoxygenic bacteria; use reduced sulfur compounds instead of water and produce sulfur granules rather than oxygen.

a. *Nitrosomonas*

b. Cyanobacteria

c. *Sphaerotilus natans*

d. Purple sulfur, or green sulfur bacteria

e. *Cytophaga*

f. *Beggiatoa*

VI. Fill-in-the-Blanks

1. Spiral and curved bacteria, as grouped in *Bergey's Manual*, are motile by means of _____.

2. The genus *Salmonella* consists almost entirely of pathogens and is taxonomically divided into hundreds of _____, rather than species.

3. *Serratia marcescens* colonies produce a _____ colored pigment.

4. A number of _____ species are plant pathogens, causing plant soft-rot diseases.

5. The cyanobacteria are often able to fix atmospheric nitrogen by means of enzymes contained in a structure called a _____.

6. Anaerobic, gram-negative, long, slender rods with pointed ends are in the genus._____ .

7. Spherical bacteria of the genus_____ occur in grapelike clusters.

8. _____ hemolytic types of bacteria form a narrow, greenish zone of hemolysis on blood agar plate.

9. _____ hemolytic types of bacteria form a clear zone of hemolysis on blood agar plates.

10. The members of the genus _____ form endospores and are obligate anaerobes.

11. *Actinomyces* resemble _____ but are obligate anaerobes rather than aerobes.

12. After an elementary body of a *Chlamydia* penetrates a host cell, it becomes a(n) _____, which then divides successively.

13. The odor of soil is due to a gas, _____, produced by *Streptomyces* bacteria.

14. The rickettsial organism that causes Q fever is named _____.

15. Members of the genus _____ produce most of our commercial antibiotics.

16. Some appendaged aquatic organisms such as the genus _____ have stalks by which they anchor themselves to surfaces.

17. The disease leptospirosis is caused by spirochetes of the genus _____.

18. A genus of bacteria resistant to most antibiotics and capable of using unusual substrates such as pesticides as nutrients is _____.

19. Gram-positive cocci that are catalase-negative are members of the genus _____.

20. The term _____ is applied to bacteria shaped as slightly curved rods.

Key

Matching

I. 1.a 2.f 3.b 4.d 5.e 6.a
II. 1.h 2.f 3.d 4.c 5.g 6.e 7.b
III.1.b 2.g 3.c 4.e 5.f 6.a 7.h
IV. 1.b 2.e 3.f 4.d 5.a 6.g 7.i 8.h
V. 1.e 2.c 3.a 4.b 5.d

Fill-in-the-Blanks

1. flagella   2. serovars (serotypes)   3. red   4. *Erwinia*   5. heterocysts
6. *Fusobacterium*   7. *Staphylococcus*   8. alpha-   9. beta-   10. *Clostridium*
11. *Nocardia*   12. reticulate body   13. geosmin   14. *Coxiella burnetii*
15. *Streptomyces*   16. *Caulobacter*   17. *Leptospira*   18. *Pseudomonas*
19. *Streptococcus*   10. vibrios

11 *Fungi, Algae, Protozoans, and Multicellular Organisms*

FUNGI

The study of fungi is **mycology**. **Molds** are multicellular, filamentous organisms that include mildews, rusts, and smuts. **Fleshy fungi** include the mushrooms and puffballs. **Yeasts** are unicellular fungi.

Vegetative Structures

Fungal filaments are called **hyphae**, and a fungal colony of hyphae is a **thallus**. Crosswalls in the hyphae are **septa**; **coenocytic hyphae** have no septa. A mass of hyphae is a **mycelium**; the **vegetative mycelium** is concerned with obtaining nutrients, and the **reproductive** or **aerial mycelium** is involved in reproduction.

Yeasts usually reproduce by **budding**, in which a protuberance on the cell enlarges and eventually separates as another cell. If daughter cells do not detach immediately, they form a short chain, a **pseudohypha**. Many yeasts are facultative anaerobes and ferment carbohydrates to product CO_2 and ethanol. Fungi that, under different conditions, grow in either yeastlike or moldlike forms, are known as **dimorphic fungi**. Features differentiating fungi from bacteria are summarized in Table 11-2 in the text.

Reproductive Structures

Asexual spores arise from one organism only. **Sexual spores** result from the fusion of nuclei from two opposite mating strains.

Asexual Spores

Arthrospores are formed by the fragmentation of a septate hypha into single cells. **Chlamydospores** form segments within a hypha. **Sporangiospores** are produced within a sac (**sporangium**) at the end of an aerial hypha called a **sporangiospore**. **Conidiospores** are produced in a chain at the end of a **conidiospore** and are not enclosed. **Blastospores** are buds.

Sexual Spores

A **zygospore** results from the fusion of the nuclei of two cells morphologically similar to each other. **Ascospores** result from the fusion of the nuclei of two cells either morphologically similar or dissimilar. These latter are produced in a sac (**ascus**). **Basidiospores** are formed externally on a base pedestal called a **basidium**. Sexual spores are the main basis for grouping the fungi taxonomically.

Nutritional Adaptations

1. Fungi usually prefer an acidic pH, too acid for most bacteria.

2. Molds are almost all aerobic.

3. Fungi are relatively resistant to osmotic pressures; e.g., they grow well at high sugar and salt concentrations.

4. Fungi are capable of growing on substances with a low moisture content on which bacteria are unable to grow.

5. Fungi require less nitrogen for equivalent weight of growth than do bacteria.

6. Fungi are capable of using complex carbohydrates such as lignins.

Medically Important Phyla of Fungi

Deuteromycota

The **Deuteromycota** also are known as the **Fungi Imperfecti**, meaning no sexual spores have yet been demonstrated. Asexual reproduction may involve chlamydospores, arthrospores, conidiospores, or budding.

Zygomycota

The **Zygomycota**, or conjugation fungi, form sexual zygospores and asexual sporangiospores. Their hyphae are coenocytic.

Ascomycota

Members of the **Ascomycota**, or sac fungi, form septate hyphae.

Basidiomycota

The **Basidiomycota**, or club fungi, have separate hyphae. They form sexual basidiospores and some form asexual conidiospores. Table 11-3 in the text summarizes the characteristics of some of the fungi.

Fungal Diseases

A fungal infection is a **mycosis**. A **systemic mycosis** occurs deep within the patient in various tissues and organs. Infections usually are transmitted by inhalation and often begin in the lungs. Fungal infections just beneath the skin are **subcutaneous mycoses**. They usually result from puncture wounds and often form disfiguring subcutaneous abscesses.

Dermatophytes are fungi that infect the epidermis, hair, and nails, causing **cutaneous mycoses**. They secrete keratinase, which degrades a protein (keratin) found in hair, skin, and nails. The infections tend to spread in a circular fashion forming **tineas (ringworms)**. Transmission is by direct contact with contaminated people, animals, or inanimate objects.

Superficial mycoses are localized along hair shafts and the superficial epidermal cells. Examples are tinea nigra and white piedra. **Mucormycosis** is an opportunistic mycosis caused by *Rhizopus* and *Mucor*, primarily in patients with ketoacidosis resulting from diabetes, leukemia, or treatment with immunosuppressive drugs. **Aspergillosis** is caused by *Aspergillus* and usually occurs in individuals with lung disease or cancer.

Candidiasis is usually caused by overgrowth by *Candida albicans*. Thrush is an oral candidiasis usually in newborns; vulvovaginal candidiasis may occur during pregnancy.

Economic Effects of Fungi

All the listed nutritional adaptations are important in understanding the reasons for fungal spoilage of dry cereals, acidic foods with high sugar content, such as jelly, growth on painted walls or on shower curtains, and the importance of fungi as plant pathogens. Examples of plant fungal diseases are Dutch elm disease and the chestnut blight; the latter essentially eliminated chestnut trees in the United States.

ALGAE

Algae are photosynthetic autotrophs; that is, they use light to convert atmospheric carbon dioxide into carbohydrates for energy. Oxygen is a byproduct of photosynthesis. **Dinoflagellates** are unicellular algae, referred to as **planktonic** because they are free-floating. Some produce neurotoxins, which may be ingested by shellfish and eventually poison humans. **Euglenoids** are unicellular, flagellated algae that are facultative chemoheterotrophs. They resemble protozoans. **Diatoms** are unicellular or filamentous algae with walls of pectin or silica that fit together as do the two halves of a Petri plate. **Brown algae**, or **kelp**, my be 50 meters in length. *Algin*, a product used as a food thickener and in the production of other goods, is extracted from their cell walls. **Red algae** are branched and multicellular, and live at greater depths than other algae. **Green algae** are usually classified as plants.

Structure and Reproduction

The body of a multicellular alga is a **thallus**, which may function as **holdfasts** to anchor them. They often have stemlike **stipes**, and leaflike **blades** (Figure 11-11c in the text). Algae are buoyed by gas-filled **bladders**.

Asexual reproduction in multicellular or filamentous algae is accomplished by fragmentation of the thallus. Following mitosis, unicellular algae divide by simple fission. Some algae reproduce sexually. **Agar** and **carrageen** are gelatinous thickeners derived from red algae.

Roles of Algae in Nature

Algae are important to the aquatic food chain because they convert carbon dioxide into consumable organic molecules. A large increase in numbers of planktonic algae is called a *bloom*. Blooms of dinoflagellates cause *red tides* in the ocean. When large numbers of algae die and decompose, oxygen dissolved in water is depleted.

LICHENS

A **lichen** is a combination of a green alga (or cyanobacterium) and a fungus. This is a *mutualistic* relationship, from which each member benefits. **Crustose** lichens grow flush onto a surface; **foliose** lichens are leaflike in shape; **fruticose** lichens have fingerlike projections. The photosynthetic alga provides carbohydrates, and the fungus provides protection from desiccation and a means of attachment.

Slime Molds

Slime molds have both fungal and animal characteristics and are classified as protists. **Cellular slime molds** resemble amoebas at one stage. When conditions are unfavorable for growth, large numbers of amoeboid cells aggregate to form a single structure, a **slug**. Cyclic AMP produced by some amoebas is the attractant towards which they migrate to form the slug. The slug moves toward light and eventually forms a stalked structure with a spore cap at the top. Under favorable conditions this spore cap differentiates into single amoeba-like spores which repeat the cycle.

    Plasmodial (acellular) slime molds are a mass of protoplasm called a plasmodium. The entire plasmodium moves like a giant amoeba and engulfs organic debris and bacteria. **Cytoplasmic streaming**, which apparently distributes oxygen and nutrients, can be observed in these slime molds. When conditions are unfavorable for growth, the plasmodium separates and stalked sporangia with spores are formed. When conditions improve, the spores germinate and the plasmodium is again formed.

PROTOZOANS

Protozoans are one-celled eucaryotic organisms that mostly feed on bacteria and small particulate nutrients. They are classified largely on the basis of their means of motility.

Protozoan Biology

Nutrition

Most protozoans are aerobic heterotrophs, but most intestinal protozoans are capable of anaerobic metabolism. A few, like Euglena, contain chlorophyll and are photoautotrophs. Some protozoans absorb food through the cell membrane, but many have a coating, the **pellicle**, that does not allow them to do so. Ciliates wave cilia (other than those used for motility) to bring food to an opening called the **cytosome**. Amoebas engulf food and phagocytize it. Digestion in all protozoans takes place in **vacuoles**, and waste is eliminated through an **anal pore**.

Reproduction

Protozoans reproduce asexually by fission, budding, or **schizogony**. The last involves multiple fission of the nucleus prior to cell division. Daughter cells eventually form around each nucleus. Sexual reproduction such as **conjugation** has been observed in some ciliates. Two cells fuse and a haploid nucleus from each migrates to the other cell. Both are now fertilized and produce daughter cells. Other protozoans produce gametes—haploid sex cells that fuse to form a zygote.

Encystment

Some protozoans produce a protective, hardened capsule called a **cyst**.

Medically Important Phyla of Protozoans

Sarcodina

Sarcodina comprise the amoebae, which move by extending lobelike projections of the cytoplasm called **pseudopods**. An important disease is amoebic dysentery.

Mastigophora

Members of the phylum **Mastigophora** move by means of sinuous movements of flagella. Some have an undulating membrane, which is perhaps a modified flagellum. The vegetative form of these protozoans is called a **trophozoite**. Representative diseases caused by this protozoan group are giardiasis, Chagas' disease, African sleeping sickness, and trichomoniasis.

Ciliata

Cilia, the organs of propulsion for the phylum ciliata, are shorter than flagella and are arranged over the cell, moving in synchrony to propel it through the medium. A dysentery caused by *Balantium coli* is the only ciliate-caused disease.

Sporozoa

Sporozoans are not motile in their mature forms and are obligate intracellular parasites. An example is the genus *Plasmodium*, which causes malaria. *Plasmodium* has a complex life cycle involving an animal host and a mosquito. The sexual cycle takes place in the mosquito, resulting in **sporozoites** that enter a host by means of mosquito bites. There they enter the liver cells, undergo **schizogony**, and leave the liver cells as **merozoites**. These merozoites infect red blood cells, again undergo schizogony and are released as more merozoites when the red blood cells rupture. This release is periodic and causes the fever and chills of malaria. Some merozoites develop into sexual forms called **gametocytes**. These are picked up by feeding *Anopheles* mosquitoes, where a new sexual cycle begins.

Toxoplasma gondii is another pathogenic sporozoan. It reproduces sexually only in cats. The feces of infected cats contain oocysts which, if ingested, infect humans or animals. This disease is dangerous to pregnant women, as it can cause congenital infections of the fetus. Table 11-6 in the text summarizes some of the parasitic protozoans.

HELMINTHS

Helminth Biology

Helminths are multicellular eucaryotic animals. Many are parasitic.

Reproduction

Adult helminths may be **dioecious**; that is, there are male and female individuals. Some, however, are **hermaphroditic**; one animal has both male and female reproductive organs.

Life Cycle

The **definitive host** harbors the adult, sexually mature helminth. **Intermediate hosts** may be necessary for larval or developmental stages.

Platyhelminthes

The **platyhelminthes**, or flatworms, are dorsoventrally flattened. The digestive system is incomplete; there is only one opening for entry of food and exit of waste. Trematodes and cestodes are members of this group.

Trematodes

Trematodes, or flukes, have flat, leaf-shaped bodies with a ventral sucker and an oral sucker to hold them in place and suck fluids from the host. They also may obtain food by absorption through the outer covering, the cuticle. A typical life cycle is that of the lung fluke. It begins with the excretion of an egg into water. A **miracidial larva** develops from the egg and enters a snail. There it produces **rediae**, each of which develops into a **cercaria**, which, in turn, penetrates a crayfish. There it encysts as a **metacercaria** in the muscles. Humans are infected by ingesting undercooked crayfish. Cercariae of diseases such as schistosomiasis are not ingested but burrow through the skin of the human host.

Cestodes

Cestodes, or tapeworms, are intestinal parasites. The head or **scolex** has suckers and usually attachment hooks. The worm consists of segments called **proglottids**, which contain both male and female organs. Mature proglottids contain fertilized eggs that are shed in the feces. The egg-laden proglottids of the beef tapeworm, for example, are ingested by grazing animals. The eggs hatch in the intestine, releasing larvae that migrate to muscles where the larval form is encysted as **cysticerci**. These may be ingested by humans, thus initiating an infestation. In some dog and cat tapeworms, humans are only intermediate hosts. The eggs are picked up from contamination of the hands by dog feces or a dog's tongue. The cyst forming in these infections of the host tissue, usually the lung or liver, is called a **hydatid cyst**.

Aschelminthes

Aschelminthes, roundworms, are cylindrical and tapered at the ends. They have a complete digestive system of mouth, intestine, and anus. Human parasites are found in the nematode group.

Nematodes

Parasitic **nematodes** (many are free-living in soil and water) do not have the succession of larval stages found in flatworms. In some cases, the eggs are infective for humans. The adults, which may be 30 cm long in the case of *Ascaris lumbricoides*, live in the intestines on semidigested food, and excrete eggs in the feces. Larvae may also be in the infective form; hookworms are an example. Hookworm larvae in the soil penetrate the human skin on contact, eventually reaching the intestines. Eggs shed in feces become larvae in the soil and continue the cycle. In trichinosis, the encysted larvae are ingested in meats such as pork. In the human host, the cysts mature into adults in the intestine, and eventually eggs from the adults become larvae and encyst in human muscles. A summary of representative parasitic helminths is presented in Table 11-7 in the text.

ARTHROPODS

Arthropods are jointed-legged animals and include the **Arachnida**, which have eight legs (spiders, mites, ticks); **Crustacea** (crabs, crayfish); **Chilopoda** (centipedes); **Diplopoda** (millipedes); and **Insecta**, which have six legs (bees, flies). Ticks and mites (Arachnida) and lice, fleas, and mosquitoes (Insecta) are particularly important in medicine as vectors of disease; that is, they carry and transmit disease-causing microorganisms.

Test

In the matching section, there is only one answer to each question; however, the lettered options (a, b, c, etc.) may be used more than once or, perhaps, not at all.

I. Match

1. The sac produced at the end of an aerial hypha called a sporangiophore.

2. Formed by fragmentation of hypha.

3. Buds.

4. Produced in a chain; not enclosed; asexual.

5. Formed on a base pedestal called a basidium.

6. Sexual spore resulting from the fusion of nuclei of two cells, morphologically either similar or dissimilar; produced in a sac.

7. Results from fusion of the nuclei of two cells morphologically similar to each other.

a. Arthrospores
b. Blastospores
c. Conidiophores
d. Conidiospores
e. Chlamydospores
f. Sporangiospores
g. Sporangium
h. Basidiospores
i. Zygospores
j. Acospores

II. Match

1. Tapeworm.

2. Roundworm.

3. Scolex.

4. Proglottids.

5. Hydatid cysts.

6. Fluke.

a. Trematode

b. Nematode

c. Cestode

III. Match

1. A resistant capsule formed by a protozoan.

2. Organelles of movement by amoebae.

3. Organism (i.e., mosquito) that transmits a disease-causing organism

4. A term describing helminths with both male and female individuals.

5. An outer covering on some helminths.

6. Outer covering on some protozoans.

a. Dioecious

b. Cuticle

c. Hermaphroditic

d. Pellicle

e. Vector

f. Pseudopods

g. Ergot

h. Sclerotia

i. Cyst

j. Schizogony

IV. Match

1. Nonmotile in mature forms.

2. Amoeba.

3. *Balantium coli*.

4. The cause of malaria.

a. Ciliata

b. Mastigophora

c. Sarcodina

d. Sporozoa

V. Match

1. Walls of pectin or silica fit together much like a Petri plate.

2. May be 50 meters in length; algin extracted from them.

3. Agar is a product of this type of organism.

4. Crustose, foliose, frutiose.

5. Cyclic AMP involved in aggregation of individual cells.

6. Composed of a mass called a plasmodium.

7. Some species responsible for red tides in the ocean.

8. A flagellated alga sometimes classified as a form of protozoan.

a. Dinoflagellates

b. Cellular slime molds

c. Kelp

d. Red algae

e. Plasmodial (acellular) slime molds

f. Diatoms

g. Euglenoids

h. Lichens

VI. Fill-in-the-Blanks

1. In the life cycle of the protozoan causing malaria, the organism grows in the red blood cell, which eventually ruptures and releases the form called _____. Then, in the mosquito the male and female gametocytes of the protozoan mate and produce _____, which the mosquito transmits to a new human host.

2. The _____, or conjugation fungi, form sexual zygospores and asexual conidiospores.

3. A fungal infection is called a _____.

4. The scientific name of the club fungi is _____.

5. _____ are fungi that infect the epidermis, hair, and nails.

6. Fungi that are sometimes yeastlike and other times filamentous are called _____.

7. _____ is the name of the opportunistic disease caused by overgrowth by *Candida albicans*.

8. The _____ are also known as the Fungi Imperfecti, meaning no sexual spores have yet been demonstrated.

9. The body of a multicellular alga is called a_____.

10. Insects have _____ legs.

11. _____, a larval form of the disease organism causing schistosomiasis, are not ingested but burrow through the skin of the human host.

12. The _____ host harbors the adult, sexually mature helminth. The _____ hosts may be necessary for larval or developmental stages.

13. An animal with both male and female reproductive organs is called _____.

14. The vegetative form of the flagellated protozoans is called a _____.

15. Fungal filaments are called _____, and a mass of hyphae is called _____.

16. _____ are unicellular fungi.

17. Hyphae with no septa are termed _____ hypae.

18. The sac in which an ascospore is formed is an _____.

19. Fungal infections just beneath the skin, usually resulting from a puncture wound, are called _____ mycoses.

20. The common name for oral candidiasis is _____.

21. Digestion in all protozoans takes place in _____ and wastes are eliminated through an anal pore.

22. One form of division by protozoan involves repeated fission of nuclei prior to cell division. This is called _____.

23. A form of sexual reproduction called _____ is found in some ciliated protozoans and involves two cells fusing together and exchanging haploid nuclei.

24. In the life cycle of the lung fluke, the rediae in the snail host develops into a _____, which in turn penetrates a crayfish.

25. The stemlike structures of a multicellular alga are called _____.

26. A lichen is a _____ type association between an alga and a fungus.

27. The structures that buoy an alga in water are called _____.

Key

Matching

I.   1.g 2.a 3.b 4.d 5.h 6.j 7.i
II. 1.c 2.b 3.c 4.c 5.c 6.a
III. 1.i 2.f 3.e 4.a 5.b 6.d
IV. 1.d ·2.c 3.a 4.d
V.   1.f 2.c 3.d 4.h 5.b 6.e 7.a 8.g

Fill-in-the-Blanks

1. merozoites; sporozoites   2. Zygomycota   3. mycosis   4. Basidiomycota
5. dermatophytes   6. dimorphic   7. candidiasis   8. Deuteromycota   9. thallus
10. six   11. cercaria   12. definitive; intermediate   13. hermaphroditic
14. trophozoite   15. hyphae; mycelium   16. yeasts   17. coenocytic   18. ascus
19. subcutaneous   20. thrush   21. vacuoles   22. schizogony   23. conjugation
24. cercaria   25. stipes   26. mutualistic   27. bladders

12 *Viruses*

In the early days of microbiology, the term **filterable agents** or **filterable virus** (the word virus derives from poison or venom) was used to designate an infectious agent that passed through filters that retained bacteria. Later the term **virus** alone came into use. At the time, no one was sure of the particulate nature of these submicroscopic agents.

General Characteristics of Viruses

Viruses are **obligatory intracellular parasites** that require a living host cell in order to multiply. The term **host range** refers to the spectrum of host cells the virus can infect. Most viruses are much smaller than bacteria, although some larger ones approach the size of very small bacteria. The size ranges from about 20 to 300 nm (see Figure 12.1 in the text).

Viral Structure

A **virion** is a fully developed complete viral particle.

Nucleic Acid

Viral nucleic acid may be either DNA or RNA in double-stranded or single-stranded forms. It may be linear, or even in several separate molecules.

Capsid and Envelope

The protein coat is the **capsid**; it is made up of protein subunits, the **capsomeres**. The capsid may be covered by an **envelope** of some combination of lipids, proteins, and carbohydrates. Envelopes may be covered with spikes projecting from the surface. Some viruses use these spikes to adhere to red blood cells, causing a clumping called **hemagglutination**. Viruses not covered by an envelope are known as **naked viruses**, or nonenveloped viruses.

General Morphology

Viruses may be classified into several morphological types. **Helical viruses** resemble long rods, their capsids a hollow cylinder with a helical structure. Examples are the tobacco mosaic virus or bacteriophage M13. **Polyhedral viruses** usually have a capsid in the shape of an **icosahedron** (a polyhedron of 20 regular triangular faces). Examples are the adenovirus and poliovirus. **Enveloped viruses** have their capsid covered by an envelope. They are roughly spherical but pleomorphic (variable in shape). A helical virus (in this case the helical capsid is folded, not extended in rodlike form) such as the influenza virus is referred to as **enveloped helical**. A polyhedral virus such as herpes simplex, with a capsule, is an **enveloped polyhedral**. **Complex viruses**, such as the poxviruses, do not contain identifiable capsids. They may have several coats around the nucleic acid or, like many bacteriophages, have a polyhedral head and a helical tail.

Classification of Viruses

In this text we group viruses according to host range; that is, animal, bacterial, or plant viruses. Present classification systems are based on type of nucleic acid, morphology, presence or absence of an envelope, etc. (See Table 12-1).

Table 12-1. Classification of Animal Viruses by Nucleic Acid Replication

Nucleic Acid Category	Viral Group and Specific Examples	Morphological Class	Dimensions of Capsid (Diameter in nm)	Clinical or Special Features
Single-stranded DNA viruses	Parvoviruses (adenosatellite)	Naked polyhedral	18-25	Very small viruses, most of which depend on coinfection with adenoviruses for growth; probably infect only rats, mice, and hamsters.
Double-stranded DNA viruses	Papovaviruses (papilloma, polyoma, simian)	Naked polyhedral	40-57	Small viruses that induce tumors; the human wart virus (papilloma) and certain viruses thatproduce cancer in animals (polyoma and simian) belong to this family. Refer to Chapters 20 and 25.
	Adenoviruses	Naked polyhedral	70-80	Medium-sized viruses that cause various respiratoryinfections in humans; some cause tumors in animals.
	Herpesviruses (herpes simplex, herpes zoster)	Enveloped polyhedral	150-250	Medium-sized viruses that cause various human diseases, such as fever blisters, chickenpox, shingles, and infectious mononucleosis; implicated in a type of human cancer called Burkitt's lymphoma. Refer to Chapters 20, 22, and 25.
	Poxviruses (variola, cowpox, vaccinia)	Enveloped complex	200-350	Very large, complex, brick-shaped viruses that cause diseases such as smallpox (variola), molluscum contagiosum (wartlike skin lesion), cowpox, and vaccinia; vaccinia virus gives immunity to smallpox. Refer to Chapter 20.
Sense strand RNA viruses	Picornaviruses (poliovirus, rhinovirus, hepatitis A virus)	Naked polyhedral	18-38	Smallest RNA- containing viruses; at least 70 human enteroviruses are known, including the polio-, coxsackie-, and echoviruses; more than 100 rhinoviruses exist and are the most common cause of colds. Refer to Chapters 21, 22, 23, and 24.

(Continued)

Table 12-1 *(Continued)*

Nucleic Acid Category	Viral Group and Specific Examples	Morphological Class	Dimensions of Capsid (Diameter in nm)	Clinical or Special Features
Sense strand RNA viruses (cont.)	Togaviruses (arboviruses, alpha virus, flavivirus)	Enveloped polyhedral	40-60	Included are many viruses transmitted by arthropods; diseases include eastern equine encephalitis (EEE), Venezuelan equine encephalitis (VEE), St. Louis encephalitis (SLE), yellow fever, and dengue. Refer to Chapters 21 and 22.
Antisense strand RNA viruses	Orthomyxoviruses (influenza A, B, C)	Enveloped helical	80-200	Medium-sized viruses with a spiked envelope; have the ability to agglutinate red blood cells; cause influenza. Refer to Chapter 23.
	Paramyxoviruses (measles, mumps)	Enveloped helical	150-300	Morphologically similar to orthomyxoviruses, but generally larger; cause parainfluenza, measles, mumps. Refer to Chapters 20, 23, and 24.
	Coronaviruses	Enveloped helical	80-130	Associated with upper respiratory tract infections and the common cold. Refer to Chapter 23.
	Rhabdoviruses (rabies)	Enveloped helical	70-180	Bullet-shaped viruses with a spiked envelope; cause rabies and Newcastle disease of chickens. Refer to Chapter 21.
	Arenaviruses (lassa)	Enveloped helical	50-300	Viruses contain RNA-containing granules; some members produce slow viral infections.
Double-stranded RNA viruses	Reoviruses	Naked polyhedral	60-80	Involved in mild respiratory infections and infantile gastroenteritis; causes Colorado tick fever.
Reverse transcription viruses	Retroviruses	Enveloped helical	100-120	Include all RNA tumor viruses. Double-stranded RNA viruses; cause leukemia and tumors in animals; cause AIDS. Refer to Chapter 18.
	Hepatitis B	Enveloped polyhedral	42	Double-stranded DNA virus; cause hepatitis; cause live tumors. Refer to Chapter 24.

Isolation, Cultivation, and Identification of Viruses

Growth of Bacteriophages in the Laboratory

Bacteriophages require a specific host bacterium for growth. The growth medium for the host may be liquid or solid, but solid media are used for detection and counting of viruses by the **plaque method**. For this method, a melted agar suspension of host cells and bacteriophage (**phage**, for short) are poured in a thin layer over an agar surface on a Petri plate. The bacteria develop into a turbid lawn except where they are destroyed by proliferating phage, forming a circular clearing called a **plaque**. Counts of phage are made in terms of **plaque-forming units**.

Growth of Animal Viruses in the Laboratory

Viruses can be cultured in suitable living animals, and some can be grown only in this way. Because signs of disease in the animal are often significant, this method is often used in diagnosis. **Embryonated eggs** can be inoculated by a hole drilled in the shell. Growth may be detected by death of the embryo or formation of pocks or lesions on the membranes. Most recently, **cell culture** (sometimes called tissue culture) has been the method of choice for viral cultivation. Animal (or plant) cells may be separated and grown as homogeneous collections of cells, not unlike bacterial cultures. In containers, the cells tend to adhere to surfaces and form a **monolayer** of cells. Cell infection by a virus causes observable death or damage known as **cytopathic effects** (**CPE**), which can be used, much as plaques are, for counting or detecting viruses.

    Primary cell lines are derived directly from tissue, and tend to die after a few generations, but a few specialized human cell lines may be cultivated for a hundred generations or so. **Continuous cell lines**, often cancer cells such as the HeLa cells, can be maintained an indefinite number of generations. **Diploid cell lines**, developed for embryonic human cells, are used to culture rabies virus for human diploid cell vaccines.

    Identification of viruses is difficult. Immunological methods using specific antibodies are most commonly used. They can be seen only with an electron microscope.

Viral Multiplication

The virus has only a few genes. Most enzymes encoded in viral nucleic are not part of the virion but are synthesized and function only within the host cell. The **viral enzymes** (or **virus-specific enzymes**) mostly replicate viral nucleic acid and seldom the viral proteins. Thus, the virus must invade a host cell and take over its metabolic machinery.

Multiplication of Bacteriophages

The most familiar example of a viral life cycle is that of the T-even phages. The tail of the phage is **adsorbed** or **attached** to the host cell. This is a highly specific reaction depending on a complementary receptor site. The phage forms a hole in the cell wall using phage lysozyme and drives the tail core through the cell wall (**penetration**); it then injects the DNA of the virus into the cytoplasm. The head (capsid) remains outside. The viral DNA causes transcription of RNA from viral DNA, and thus commandeers the metabolic machinery of the host cell for its own biosynthesis. For several minutes following infection, complete phages cannot be found; this is called the eclipse period. During the **maturation period**, which follows, the phage DNA and capsid, formed separately, are assembled into virions. When complete, the host cell **lyses** and **releases** these virions. The time required from phage adsorption to release is the **burst time** (about 20-40 minutes), and the number released is the burst size (about 50-200). The stages of phage multiplication can be demonstrated with the **one-step growth experiment** (see Figure 12-15 in the text).

Lysogeny. Sometimes the lytic cycle just described does not occur. The phage DNA becomes incorporated as a prophage into the host's DNA, a state called **lysogeny**. Such phages are **lysogenic** or **temperate** phages and their bacterial host, **lysogenic cells**. The lysogenized cell may exhibit new properties such as toxin production (examples are scarlet fever, diphtheria, and botulism). The prophage is reproduced along with the bacterial chromosome but can be induced to complete the lytic cycle. In **specialized transduction** a lysogenic phage incorporates small amounts of host DNA and can confer this DNA onto a newly infected cell.

Multiplication of Animal Viruses

Multiplication of animal viruses follows the general pattern just described, but with important differences. Animal viruses have no tail, so adsorption is by spikes, small fibers, etc. Penetration occurs by fusion of the envelope with the host plasma membrane or by the naked virus somehow entering the cytoplasm. Penetration by phagocytosis is called **endocytosis**. An alternative method of penetration is **fusion**. The viral envelope fuses with the plasma membrane and releases the capsid into the host cell's cytoplasm. Once inside the host cell, the viral nucleic acid separates from the protein coat, a process called **uncoating**.

Biosynthesis for DNA-containing viruses. Multiplication of DNA viruses may occur entirely in the cytoplasm (poxviruses). Or, the DNA may be formed in the nucleus and the protein in the cytoplasm, final assembly taking place in the nucleus (parvoviruses, papovaviruses, adenoviruses, herpesviruses).

Biosynthesis for RNA-containing viruses. Multiplication of RNA viruses essentially is similar to that of DNA viruses. However, these viruses must cause formation of an **RNA-dependent RNA polymerase enzyme** in order to make an RNA strand complementary to that of the infecting RNA. There are several paths by which this is accomplished.

Picornavirus. These are single-stranded RNA viruses such as poliovirus. The single strand (+ or **sense strand**) acts as mRNA. It serves as a means to make RNA-dependent RNA polymerase, which catalyzes the synthesis of another strand of RNA (- or **antisense strand**). The - strand serves as a template for + strands that in turn serve as a means to produce viral RNA or viral protein.

Togavirus. Also containing a single + strand of RNA, togaviruses differ from picornaviruses in that two types of mRNA are transcribed from the - strand. One codes for capsid proteins and the other for envelope proteins.

Rhabdovirus. These contain a single - strand of RNA. Since rhabdoviruses already contain RNA-dependent RNA polymerase they do not have to synthesize this enzyme. The RNA polymerase produces a + strand, which serves as mRNA and a template for synthesis of viral RNA.

Reovirus. These contain double stranded RNA. One of the capsid proteins of these viruses serves as RNA-dependent RNA polymerase. After the capsid enters a host cell, mRNA is produced inside the capsid and released into the cytoplasm. There it is used to synthesize more viral proteins. One of these proteins acts as RNA-dependent RNA polymerase to produce - strands of RNA. These - strands and the mRNA + strands form the double-stranded RNA in reoviruses.

Retroviruses. Some retroviruses cause cancers, and one type has been implicated as the cause of **acquired immune deficiency syndrome (AIDS)**. These viruses carry a polymerase (**reverse transcriptase**) that uses the RNA of the virus to make a complementary strand of DNA (hence reverse transcription, or **retrovirus**). This DNA becomes integrated into the DNA of a host cell (**provirus**), and transcription into mRNA may then take place normally.

Maturation and release. For enveloped cells, the envelope develops around the capsid from the plasma membrane by a process called **budding**, which occurs as the nucleic acid enclosed in the capsid pushes out through the plasma membrane. Budding does not necessarily kill the host cell. Naked viruses released by host cells usually cause **lysis** and death of the host cell.

Effects of Animal Viral Infection on Host Cells

Abnormalities resulting in the damage or death of a host cell are **cytopathic effects** (CPE). One type of CPE is an **inclusion body**; examples are the Negri bodies in rabies-infected cells. These are probably accumulations of viruses or viral parts. Some host cells fuse to form giant cells called **polykaryocytes**. **Interferon** may be produced by viral infected cells to help protect neighboring cells from infection.

Viruses and Cancer

When cells multiply in an uncontrolled way, the excess tissue is called a tumor, which is **malignant** if cancerous and **benign** if not. Spread of cancer cells in the body is termed **metastasis**. **Leukemias** are not solid tumors but an excess production of white cells. Chicken **sarcoma** (cancer of connective tissue) and **adenocarcinoma** (cancer of glandular tissue) can be transmitted by viruses

Transformation of Normal Cells into Tumor Cells

It is believed that cancer-causing changes in cellular DNA are directed by parts of the genome called **oncogenes**. Under normal conditions oncogenes probably code for necessary cellular proteins (**proto-oncogenes**), but mutations can cause oncogenes to bring about cancerous transformations of cells. Oncogenes can be activated by chemicals, oncogenic viruses, and radiation. A tumor cell formed by activation of an oncogene is said to be **transformed** and is distinctly different from normal cells. Sometimes the provirus remains latent and replicates only with the host cell, or it may become transcribed and produce new, infective viruses. Finally, it may convert the host cell into a tumor cell. Transformed cells lose contact inhibition, that is, they do not stop reproduction when in contact with neighbor cells. They also contain **tumor specific transplantation antigens** (TSTA) on the surface, or **T antigens** in the nucleus.

DNA-Containing Oncogenic Viruses

Adenoviruses, herpesviruses, poxviruses, and papovaviruses groups all contain **oncogenic viruses**. Among the papovaviruses are the papilloma viruses that cause warts and the polyoma viruses that cause several types of mouse tumors. Viral DNA is integrated into the host cell DNA. Herpesviruses, including the Epstein-Barr (EB) virus, are attracted to lymphocytes (a type of white blood cell). There, the EB virus may cause Burkitt's lymphoma or nasopharyngeal carcinoma.

RNA-Containing Oncogenic Viruses

Only the retroviruses seem to be oncogenic among the RNA types. Their oncogenic activity seems related to the production of reverse transcriptase. The DNA synthesized from viral RNA becomes integrated into the host cell's DNA (provirus). In some cases, this may convert the host cell into a tumor cell.

Activation of Oncogenes

Cancer appears to be a multistep process in which at least two oncogenes must be activated. Four mechanisms for activation have been proposed.

1. **A single mutation** alters the amino-acid sequence of the gene product, possibly affecting regulation of the oncogene. A possible result is uncontrolled production of a growth factor, and uncontrolled cellular growth.

2. **Transduction** of oncogenes by a virus could remove the oncogenes from normal cellular controls. The viral regulatory proteins could cause abnormal formation of oncogene products.

3. **Translocation** of an oncogene could place it at a position on the chromosome where normal controls are not active.

4. **Gene amplification** could cause the gene to be replicated several times, which causes the cell to produce unusually large amounts of oncogenic products.

Latent Viral Infections

Sometimes the virus remains latent in the nerve cells of the host for long periods without causing disease. Stress or some other cause may trigger their reappearance. This is the case with the herpes simplex virus, which causes cold sores, and the chickenpox virus, which causes shingles.

Unconventional Agents of Disease

The term **slow viral infections** usually is reserved now for diseases that apparently are not caused by conventional viruses. The prototype disease is scrapie in sheep. Hypotheses concerning the nature of these unconventional agents are:

1. The agent may be a conventional virus but the nucleic acid is extraordinarily difficult to detect.

2. The agent might contain an undetectably small nucleic acid enclosed in protein formed by the host. The nucleic acid may not encode protein but cause disease by disrupting the host cell. This hypothetical agent has been called a **virino**.

3. The infectious agent might be pure protein with an unknown mode of replication that does not involve nucleic acid. This postulated agent has been named a **prion**.

Diseases caused by these agents, other than scrapie in sheep, included Creutzfeldt-Jakob disease, kuru, and Gerstmann-Straussler syndrome.

Plant Viruses and Viroids

Some plant diseases are caused by **viroids**. These are very short pieces of naked RNA with no protein coat.

Test

In the matching section, there is only one answer to each question; however, the lettered options (a, b, c, etc.) may be used more than once or, perhaps, not at all.

I. Match

1. A complete, assembled virus.

2. The subunits making up the protein outer coating of most viruses.

3. The protein outer coating of most viruses.

4. The term once meant poison or venom.

5. A combination of lipids, proteins, and carbohydrates covering the protein coating of a virus.

a. Virion
b. Capsid
c. Capsomere
d. Envelope
e. Virus

II. Match

1. Describes the morphology of the capsid of many viruses.

2. A method by which a virus enters an animal host cell.

3. A cell line derived from tissue that normally reproduces for relatively few generations.

4. The HeLa cell line would be placed in this group.

5. A clearing in a "lawn" of susceptible bacterial cells.

6. The number of bacteriophages produced by one bacterial host cell.

7. An unconventional agent.

8. A bacterial virus.

9. A short strand of RNA virus without a capsid.

a. Burst size
b. Burst time
c. Primary cell line
d. Continuous cell line
e. Plaque
f. Cytopathic effect
g. Icosahedral
h. Endocytosis
i. Phage
j. Viroid
k. Diploid cell line
l. Prion

III. Match

1. Describes a method by which an enveloped virus leaves the host cell while acquiring the envelope.

2. Describes growth characteristics of cell cultures in glass or plastic.

3. A term meaning "cancer causing."

4. Observable changes in a virus-infected cell.

5. The time during which the capsids and DNA of a phage, already formed, are now assembled into complete viruses.

a. Replicative form
b. Maturation period
c. Budding
d. Oncogenic
e. Cytopathic effect
f. Endocytosis
g. Monolayer
h. Eclipse period

IV. Match

1. Giant cells formed as a cytopathic effect in cell cultures.

2. Clumping of red blood cells due to adherence to spikes on viruses.

3. Equivalent to mRNA in a single strand RNA virus.

4. RNA to DNA.

a. Polykaryocytes

b. + or sense strand

c. Reverse transcription

d. Interferon

e. Hemagglutination

V. Fill-in-the-Blanks

1. The virus, once inside the host cell, separates the viral nucleic acid from the capsid, this is called _____.

2. The cytophathic effect known as a Negri body in a rabies-infected cell is an example of a(n) _____ body.

3. _____ are not solid tumors, but an excessive production of white blood cells.

4. Spread of cancer cells in the body is called _____.

5. The herpes simplex virus remains_____ in nerve cells of the host for long periods without causing disease.

6. Counts of phage are made in terms of _____ units.

7. Viruses not covered by an envelope are known as _____viruses.

8. The term _____ refers to the spectrum of host cells the virus can infect.

9. When cells multiply in an uncontrolled way, the excess tissue is called a _____.

10. Oncogenic viruses are those that _____ cells into tumor cells.

11. The type of virus implicated as a cause of AIDS is a _____.

12. The acronym TSTA stands for tumor specific _____ antigens.

13. For several minutes following infection by a phage, no complete phages can be found in the host cell; this is called the _____ period.

14. The _____ of the phage is adsorbed to the host cell.

15. The phage forms a hole in the cell wall using phage _____ and drives the tail core through the cell wall.

16. Sometimes the lytic cycle does not occur upon phage infection of a host bacterium. The phage DNA becomes incorporated as a _____ into the host's DNA.

17. When the phage DNA is incorporated into the host's DNA, this state is called _____.

18. Transformed cells lose _____; that is, they do not stop reproduction when in contact with neighbor cells.

19. An unconventional agent hypothesized to contain an undetectably small amount of nucleic acid surrounded by protein formed by the host is a _____.

20. _____ may be produced by viral-infected cells to help protect neighboring cells from infection.

21. Tumors are malignant when cancerous, and _____ when not cancerous.

Key

Matching

I. 1.a 2.c 3.b 4.e 5.d
II. 1.g 2.h 3.c 4.d 5.e 6.a 7.l 8.i 9.j
III. 1.c 2.g 3.d 4.e 5.b
IV. 1.a 2.e 3.b 4.c

Fill-in-the-Blanks

1. uncoating 2. inclusion 3. leukemias 4. metastasis 5. latent 6. plaque-forming 7. naked 8. host range 9. tumor 10. transform 11. retrovirus
12. transplantation 13. eclipse 14. tail 15. lysozyme 16. prophage
17. lysogeny 18. contact inhibition 19. virino 20. interferon 21. benign

13 *Principles of Disease and Epidemiology*

Pathology, Infection, and Disease

Pathology is the science that deals with the study of disease. It involves the **etiology** (cause) of the disease, the manner in which a disease develops (**pathogenesis**), the structural and functional changes brought about by the disease, and the final effects on the body. **Infection** means invasion or colonization of the body (the host in this case) by potentially pathogenic microorganisms. **Disease** itself is any change from a state of health, an abnormal state in which the body is not properly adjusted or capable of performing its normal functions.

Normal Flora

Microorganisms that establish permanent residence without producing disease are known as **normal flora** or **normal microbiota**. Other microorganisms that may be present for a time and then disappear are called **transient flora**. Normal flora can benefit the host by preventing the overgrowth of harmful microorganisms, a process called **microbial antagonism**. The relationship between the normal flora of a healthy person and that person is called **symbiosis**. If, in the symbiosis, one of the organisms is benefitted and the other unaffected, the relationship is known as **commensalism**. If both organisms are benefitted it is called **mutualism**, and if one organism is benefitted at the expense of the other, it is **parasitism**. Under certain conditions these relationships can change, and members of the normal flora can become **opportunistic pathogens**.

Etiology of Infectious Disease

Infectious diseases. Disease-producing microorganisms, **pathogens**, probably account for half of all human diseases. Examples of such diseases are gonorrhea, measles, and meningitis.

Koch's Postulates

These postulates must be fulfilled in order to demonstrate that a specific microorganism is the cause of a specific disease.

1. The same pathogen must be present in every case of the disease.

2. The pathogen must be isolated from the diseased host and grown in pure culture.

3. The pathogen from the pure culture must cause the disease when inoculated into a healthy, susceptible laboratory animal.

4. The pathogen must again be isolated from the inoculated animal and must be shown to be the same pathogen as the original organism.

A few diseases, such as syphilis, are caused by organisms that cannot be isolated and grown on laboratory media. However, evidence based on the first postulate over many years' experience has been adequate to show that a certain organism is associated with the disease. Some diseases have such clearly defined symptoms, diphtheria or tetanus, for example, that associations between specific microorganisms and the disease are obvious. Other diseases such as pneumonia, peritonitis, and meningitis may be caused by a variety of microorganisms and the specific etiology is not easy to determine from the symptoms. Furthermore, some pathogens may infect a number of different organs or tissues and cause very different diseases or disease symptoms. A good example is *Streptococcus pyogenes*, which can cause sore throats, scarlet fever, erysipelas, and puerperal fever.

Classifying Infectious Diseases

Symptoms are changes in body function felt by the patient, such as pain and malaise. These are subjective and not apparent to an observer. The patient also may exhibit **signs**, objective changes that the physician can observe and measure. A specific group of symptoms or signs accompanying a particular disease is called a **syndrome**.

Communicable diseases are spread directly or indirectly from one host to another; typhoid fever and tuberculosis are examples. **Noncommunicable diseases** are caused by microorganisms such as *Clostridium tetani*, which only produces tetanus when it is introduced into the body by contamination of wounds. A **contagious disease** is easily spread from one person to another. The **incidence** of a disease is the fraction of a population that contracts it during a particular length of time. The **prevalence** of a disease is the fraction of the population that has the disease at a given time.

If a disease occurs only occasionally, it is called **sporadic**, and when it is constantly present, as is the common cold, it is termed **endemic**. If many people in a given area acquire a certain disease in a short period of time, it is referred to as an **epidemic** disease. Influenza is an example of this also. A worldwide epidemic is referred to as a **pandemic disease**.

Severity or Duration

Acute disease is one that develops rapidly but lasts only a short time—influenza, for example. A **chronic disease** develops more slowly and the body reactions are often less severe, but it is continuous or recurrent for long periods of time. Tuberculosis, syphilis, and leprosy are such diseases. Diseases intermediate between acute and chronic are described **subacute**.

Extent of Host Involvement

A **local infection** is one in which the invading microorganisms are limited to a relatively small area of the body, as for example, in boils or abscesses. In a **systemic**, or **generalized**, infection microorganisms or their products are spread throughout the body by the blood or lymphatic system. A **focal infection** is one in which a local infection, such as infected teeth, tonsils, or sinuses, enters the blood or lymph and spreads to other parts of the body. The presence of bacteria in the blood is known as **bacteremia**, and the rapid multiplication of bacteria in the blood is called **septicema**. **Toxemia** is the presence of toxins in the blood, and **viremia** is the presence of viruses in the blood. A **primary infection** is an acute infection that causes the initial illness. A **secondary infection** is one caused by an opportunist only after the primary infection has weakened the body's defenses. An **inapparent**, or **subclinical**, **infection** is one that does not cause any noticeable illness.

Spread of Infection

Reservoirs

Human reservoirs. Many people harbor pathogens and transmit them to others, directly or indirectly. These persons may be diseased and obvious transmitters, but others, called carriers, do not exhibit symptoms.

Animal reservoirs. Diseases that occur primarily in wild and domestic animals, but that can be transmitted to humans are called **zoonoses**. Transmission may be by direct contact with infected animals; contamination of food and water; insect vectors; contact with contaminated hides, fur, or feathers; or consumption of infected animal products.

Nonliving reservoirs. Examples of nonliving reservoirs for infectious diseases are soil, which harbors the agent of botulism, and water contaminated by human or animal feces, which transmits gastrointestinal pathogens.

Transmission of Disease

Contact Transmission. Infections may be spread more or less directly from one host to another by *direct contact*, such as kissing, hand-shaking, bites, or sexual intercourse. *Droplet infection*, in which agents of disease are spread very short distances while contained in droplets of saliva or mucus from coughing or sneezing, also is considered a form of contact transmission.

Common Vehicle Transmission. Inanimate reservoirs such as food, water, or blood may transmit diseases to large numbers of individuals.

Airborne Transmission. Diseases spread by agents of infection traveling on droplets or dust for a distance of more than a meter are considered to be airborne. Spores produced by fungi also can be transmitted by the airborne route.

Vectors. Arthropods are the most important group of disease vectors; that is, animals that carry pathogens from one host to another. In **mechanical transmission**, insects, such as flies, carry pathogens on their bodies to food that is later swallowed by the host. In **biological transmission**, the arthropod may pass the pathogen in a bite, or it may pass the pathogen in its feces, which later enters the wound caused by the arthropod's bite.

Nosocomial (hospital-acquired) infections. Hospital patients often are **compromised hosts** with lowered resistance and increased susceptibility of infection. The three most likely conditions to compromise a host are 1) broken skin or mucous membrane, 2) a suppressed immune system, and 3) impaired cell activity. Many nosocomial infections are caused by opportunistic pathogens. At one time nosocomial diseases were caused mostly by gram-positive bacteria such as streptococci or staphylococci. Although antibiotic-resistant strains of these organisms are still significant causes of nosocomial diseases, today most are caused by gram-negative bacteria such as *Pseudomonas aeruginosa* or *Escherichia coli*.

Patterns of Disease

Predisposing factors such as gender, genetic background, climate, age, nutrition, and other personal factors can greatly affect the occurrence of disease in individuals.

Development of Disease

The development of disease follows a certain sequence of steps. The **period of incubation** is the time between actual infection and the first appearance of signs or symptoms. The **prodromal period** follows the incubation period in some disease and is is characterized by mild symptoms of the disease. During the **period of illness**, the overt symptoms of the disease are apparent. During the **period of decline** the signs and symptoms subside. If the decline is short, this is said to occur by **crisis**. If several days pass, this is called **lysis**. The patient regains his or her prediseased state during the **period of convalescence**.

Epidemiology

The science that deals with the transmission of diseases in the human population, and where and when they occur, is called **epidemiology**. An epidemiologist determines not only the etiology of a disease, but also data such as geographical distribution, and gender, age, etc., of persons affected. Figure 13-1 shows the graphs obtained from some hypothetical epidemiologic data. First, an epidemiologist collects data such as the place(s) where a disease occurred and time(s) when it occurred. Persons affected by the disease would provide information on age, sex, personal habits, etc. This process is known as **descriptive epidemiology**. These data are then studied (**analytical epidemiology**) to look for common factors among the affected persons that might have preceded the disease outbreak.

    Experimental epidemiology tests a hypothesis, such as assumed effectiveness of a drug. Randomly selected groups receive either one of two substances, the selected drug or a placebo (a substance that has no effect). The groups are compared to discern the effectiveness of the drug versus the placebo.

    The Centers for Disease Control (CDC), a branch of the U.S. Public Health Service, located in Atlanta, Georgia, are a central source of epidemiological information in the United States. The CDC issues the *Morbidity and Mortality Weekly Report* (*MMWR*), which contains data on morbidity (relative incidence of a disease) and mortality (deaths from a disease). **Notifiable diseases** are those for which physicians must report cases to the Public Health Service.

Figure 13.1 Epidemiological graphs.
Epidemiologists use various types of
graphs to describe patterns of disease.
(a) This graph of the number of
legionellosis cases reported from 1976 to
1987 simply shows the yearly occurrence of
the disease during that time period.
(b) This bar graph looks at legionellosis
from a different perspective, allowing
epidemiologists to begin drawing some
conclusions about the disease. It records
the incidence by month, for one year
(1979). Over a period of years, researchers
saw that the incidence was consistently
highest during the hot months. These data
helped them realize that the disease was
transmitted by an airborne pathogen
through air conditioning cooling towers in
use during the hot weather. Note that this
graph records the number of cases per
100,000 people, rather than the total
number of cases.
(c) Annual incidence of toxic shock
syndrome among menstruating women,
1979-1986. This graph shows how public
education can help in disease control.
Once women learned how tampons were
involved in the disease and decreased use
of them, incidence of the disease fell
sharply.

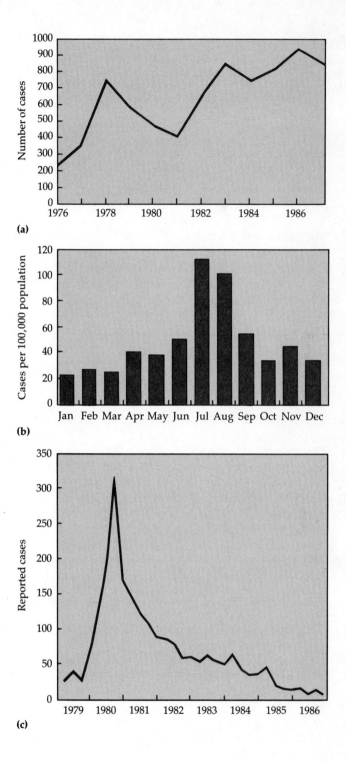

Test

In the matching section, there is only one answer to each question; however, the lettered options (a, b, c, etc.) may be used more than once or, perhaps, not at all.

I. Match

1. Invasion or colonization of the body by potentially pathogenic microorganisms.

2. The cause of the disease.

3. A change from a state of health in which the body is not properly adjusted or capable of performing its normal functions.

4. The manner in which a disease develops.

a. Infection

b. Pathogenesis

c. Disease

d. Etiology

II. Match

1. One organism is benefitted at the expense of another.

2. The general relationship between the normal flora of a healthy person and that person.

3. One of the organisms is benefitted and the other unaffected.

a. Symbiosis
b. Opportunistic
c. Commensalism
d. Mutualism
e. Parasitism

III. Match

1. First, symptoms such as headache or malaise appear.

2. Individual regains strength and body returns to its prediseased state.

3. Time between infection and first appearance of signs and symptoms.

a. Prodromal period
b. Period of convalescence
c. Period of incubation
d. Period of illness

IV. Match

1. Easily spread from one person to another person.
2. Tetanus is an example.

   a. Contagious disease

   b. Communicable disease

   c. Noncommunicable disease

V. Match

1. Inanimate object that may transmit disease.
2. A group of symptoms associated with a disease.
3. Identification of a disease.
4. Objective changes caused by a disease that the physician may observe.
5. An arthropod, for example, that carries malaria.
6. The rapid growth of a bacterium in the blood or lymph.

   a. Septicemia
   b. Bacteremia
   c. Syndrome
   d. Diagnosis
   e. Signs
   f. Vector
   g. Fomite

VI. Fill-in-the-Blanks

1. Diseases acquired in a hospital are called _____ diseases.

2. Persons who transmit diseases, but who do not exhibit any symptoms of illness are called_____.

3. In _____ transmission of disease, an insect such as a fly carries the pathogen on its body to human food.

4. Diseases that occur in animals and can be transmitted to humans are called _____.

5. The _____ of a disease is the fraction of the population that contracts it during a particular period of time.

6. _____ disease is one that develops rapidly but lasts for only a short time.

7. A disease that occurs only occasionally is called _____.

8. The simple presence of bacteria in the blood is known as _____.

9. A _____ infection is one caused by an opportunist after the primary infection has weakened the body's defenses.

10. _____ are changes in body function felt by the patient and subjective in nature, such as pain.

11. The science that deals with transmission of diseases in the human population, and when and where they occur is called _____.

12. The acronym CDC stands for_____.

13. An abscess is an example of a _____ type of infection.

14. An infection in which the microorganisms or their products are spread throughout the body in the blood or lymphatic system is known as a _____ infection.

15. An inapparent, or _____, disease is one that does not cause any noticeable illness.

16. The _____ of a disease is the fraction of the population having the disease at a given time.

17. Diseases between acute and chronic are described as being _____.

18. A worldwide epidemic is referred to as a _____ disease.

19. A pathogen is found in all cases of a certain disease and grown in pure culture; then it is inoculated into a laboratory animal. What is the next step of Koch's Postulates?

_____

_____

20. _____ is the science that deals with the study of disease.

Key

Matching

I. 1.a 2.d 3.c 4.b
II. 1.e 2.a 3.c
III. 1.a 2.b 3.c
IV. 1.a 2.c
V. 1.g 2.c 3.d 4.e 5.f 6.a

Fill-in-the-Blanks

1. nosocomial   2. carriers   3. mechanical   4. zoonoses   5. incidence   6. acute
7. sporadic   8. bacteremia   9. secondary   10. symptoms   11. epidemiology
12. Centers for Disease Control   13. local   14. systemic   15. subclinical
16. prevalence   17. subacute   18. pandemic
19. Isolate the pathogen from the animal and show that it is the same as the original organism.
20. pathology

14 *Mechanisms of Pathogenicity*

Pathogenicity is the ability to cause disease in a host. First, however, the pathogen must enter the host's body. **Virulence** is the degree of pathogenicity.

Entry of a Microorganism into the Host

The avenue by which a microbe gains access to the body is a **portal of entry**.

Mucous Membranes

To gain access to the body, pathogens can penetrate mucous membranes lining the conjunctiva, and respiratory, gastrointestinal, and genitourinary tracts. The respiratory tract is the easiest, most frequently used route of entry for infectious microorganisms. Diseases contracted by this route are the common cold, penumonia, tuberculosis, influenza, measles, and smallpox. Microorganisms contracted from food, water, or fingers enter the body by the gastrointestinal tract. Although many are destroyed by stomach acids and intestinal enzymes, diseases such as poliomyelitis, hepatitis A, typhoid fever, amoebic dysentery, bacillary dysentery, and cholera are transmitted in this manner.

Skin

With a few exceptions, such as the hookworm, microorganisms cannot penetrate unbroken skin. Some fungi, however, grow on the keratin of the skin, and other microorganisms gain access by penetrating into openings such as hair follicles and sweat ducts. Also, many bacteria and viruses can penetrate the intact mucous membranes, for example, the organism causing syphilis.

Parenteral Route

When the skin and mucous membranes are punctured or injured (traumatized), microorganisms gain access to body tissues. This route is referred to as the **parenteral route**.

Preferred Portal of Entry

Whether or not disease results after entry of microorganisms depends on many factors. The organism must enter by a preferred route. *Salmonella typhi* for example, must enter the gastrointestinal tract if it is to cause typhoid, rather than being rubbed onto the skin. *Clostridium tetani* must penetrate the skin to cause tetanus, and *Corynebacterium diphtheriae* must enter the respiratory tract if it is to cause diphtheria.

Numbers of Invading Microbes

A measure of virulence is the LD_{50} (**lethal dose**), which is the dose of pathogen that will kill half of the test animals. If the pathogen causes only a nonfatal disease, the test is referred to as ID_{50} (**infectious dose**). The fewer microorganisms required, the higher the virulence.

Adherence

For most pathogens, adherence is necessary for pathogenicity. The attachment between pathogen and host makes use of surface **ligands** (**adhesins**) and complementary surface **receptors** on the host cells. Ligands are mostly glycoproteins or lipoproteins which are frequently associated with structures such as pili; the receptors are usually sugars such as mannose.

How Pathogens Penetrate Host Defenses

Capsules. For many pathogens, such as *Streptococcus pneumoniae* and *Klebsiella pneumoniae*, **capsules** confer a resistance to phagocytosis.

Components of the Cell Wall. Cell surfaces sometimes contribute to invasiveness. *Streptococcus pyogenes*, for example, contains a protein called M protein on the fimbriae and cell surface that helps it resist phagocytosis and improve adherence.

Enzymes. Extracellular **enzymes** (exoenzymes) have the ability to break open cells, dissolve material between cells, and form or dissolve blood clots. These may contribute to invasiveness.

Leukocidins are substances produced by some bacteria that destroy certain phagocytic cells. **Hemolysins** are enzymes that cause lysis of red blood cells. Staphylococci, streptococci, and *Clostridium perfringens*, the agent of gas gangrene, are important pathogens producing hemolysins. **Coagulases** are enzymes that coagulate blood. The fibrin clot formed may protect the bacterium from phagocytosis. Coagulase is produced by some members of the genus *Staphylococcus*. Bacterial **kinases** are enzymes that break down fibrin and dissolve clots formed by the body to isolate infections. Streptokinase (fibrinolysin), which is produced by streptococci, and staphylokinase, produced by staphylococci, are the best known. **Hyaluronidase** is an enzyme secreted by certain bacteria that digests hyaluronic acid. This compound is a mucopoly saccharide that holds together certain body cells, much like mortar holds together bricks, especially in connective tissue. Both the organisms that produce gas gangrene, as well as a number of streptococci, produce this enzyme, which may promote spread of the infections to adjoining tissues. **Collagenase**, produced by several species of *Clostridium*, breaks down the collagen framework of the muscle tissue. **Necrotizing factors**, which cause the death of body cells, **hypothermic factors** affecting body temperature, **DNase** and **streptodornase** that reduce the viscosity of pus, **lecithinase** that destroys plasma membrane, and **protease** that breaks down muscle tissue also may be important to virulence.

Damage to Host Cells

Toxins

Toxins are poisonous substances produced by certain microorganisms. The capacity to produce them is called **toxigenicity**. The presence of toxins in the blood and lymph is called **toxemia**.

Exotoxins. Exotoxins are proteins released from the bacterium into the surrounding medium. Most bacteria involved are gram-positive, and the genes are probably all carried on bacterial plasmids or phages. Examples are **botulinal toxin**, which acts to prevent transmission of nerve impulses; **tetanus toxin**, which causes excitation of the central nervous system; **diphtheria toxin**, which inhibits protein synthesis in eucaryotic cells; **staphylococcal enterotoxin**, which causes diarrhea and vomiting; and *Vibrio* **enterotoxin** (**choleragen**), which causes diarrhea by altering the host's water and electrolyte balance. The body produces antibodies called **antitoxins**, by heat, formaldehyde, and other means. These are called **toxoids**; they stimulate immunity to the toxin when injected into the body. When grouped by mode of actions: 1) **cytotoxins** kill host cells, 2) **neurotoxins** affect the nervous system, and 3) **enterotoxins** affect the gastrointestinal tract.

Endotoxins. Endotoxins are not secreted by the cell like the exotoxins. They are not proteins, but are lipopolysaccharide components of the cell walls of most gram-negative bacteria. They are released upon the death of the cell. They cause fever, weakness, aches, and sometimes shock. Fever related to endotoxins is triggered by **interleukin-1 (IL-1)** produced by macrophages after ingestion of bacteria. Shock is caused by **tumor necrotizing factor (cachectin)** produced by phagocytosis of gram-negative bacteria. There are no effective antitoxins, and antibodies are not very effective. Diseases associated with endotoxins are typhoid fever, bacillary dysentery, and epidemic meningitis.

Plasmids, Lysogeny, and Pathogenicity

Several diseases are caused only when the pathogen has lysogenized or carries a particular plasmid.

Pathogenic Properties of Nonbacterial Microorganisms

Viruses

Some viruses cause **cytocidal effects** (changes resulting in cell death), while other viruses are noncytocidal. Some **cytopathic effects**, or observable changes, are:

1. Cellular macromolecular synthesis may be stopped.

2. Enzymes are released from lysosomes and cause autolysis of the cell.

3. Inclusion bodies are formed, viral parts being assembled into complete virions.

4. Host cells fuse to produce "giant" cells.

5. Antigenic changes are induced on the surface of infected cells, thus making destruction of the cell by the host's immune system more likely.

6. Chromosomal changes are induced in the host cell.

7. Host cells are transformed to spindle-shaped cells that do not have contact inhibition and grow without regulation.

Fungi, Protozoans, Helminths, and Algae

Some protozoans and helminths grow on host tissues, which causes cellular damage to the host. Their waste products may contribute to symptoms of disease in the host. Some algae produce neurotoxins.

Some toxins are associated with fungi, for example **ergot**, which causes **ergotism**. Ergot is contained in sclerotia (resistant mycelia) of the plant pathogen *Claviceps purpurea*. Ergotism can result in hallucinations, or it may constrict blood capillaries and cause gangrene in the extremities. **Aflatoxin** is produced by the mold *Aspergillus flavus* and can cause cancer of the liver in animals. Toxic mushrooms such as *Amanita phalloides* contain the dangerous neurotoxins **phalloidin** and **amanitin**.

Some algae produce neurotoxins and saxitoxin. For example, people become ill after ingesting mollusks that feed on the algae.

Test

In the matching section, there is only one answer to each question; however, the lettered options (a, b, c, etc.) may be used more than once or, perhaps, not at all.

I. Match

1. Produced by some members of the genus *Staphylococcus*; forms a fibrin clot around the bacterium.

2. Substance produced by some bacteria that destroys certain phagocytic cells.

3. Enzymes that cause lysis of red blood cells.

4. Enzymes that break down fibrin and dissolve clots.

5. The fibrinolysins produced by the streptococci.

6. May cause hallucinations or gangrene.

7. Tumor necrotizing factor.

a. Leukocidins
b. Collagenase
c. Kinases
d. Hyaluronidase
e. Coagulase
f. Hemolysins
g. Cachectin
h. Aflatoxin
i. Amanitin
j. Ergot

II. Match

1. A protein secreted by the bacterium.

2. The tetanus toxin is a good example.

3. A lipopolysaccharide component of the cell wall of many gram-negative bacteria.

4. Their presence in the body causes the formation of antibodies by the host.

5. When altered to an antivirulent form and then injected to stimulate immunity in humans, they are called toxoids.

6. Released upon death of the cell.

a. exotoxin

b. endotoxin

III. Match

1. The capacity to form toxins.

2. The ability to cause disease in a host.

3. The presence of a toxin in the bloodstream.

a. Pathogenicity
b. Toxemia
c. Toxoid
d. Antitoxin
e. Toxigenicity

IV. Match

1. A measure of virulence.

2. The degree of pathogenicity.

3. Ability of a pathogen to establish residence in a host.

a. Traumatized
b. Virulence
c. Pathogenicity
d. LD_{50}
e. Invasiveness

V. Fill-in-the-Blanks

1. The route followed when the skin or mucous membrane is punctured and pathogens gain access to body tissues is called the _____ route.

2. The term LD_{50} refers to the dose of pathogen that will kill half of the test _____.

3. Capsules aid in invasiveness because they confer a resistance to _____ on many pathogens.

4. _____, produced by several species of *Clostridium*, breaks down the collagen framework of the muscle tissue.

5. Hyaluronidase is an enzyme secreted by certain bacteria that digests _____.

6. Diseases associated with _____-toxins are typhoid fever, bacillary dysentery, and epidemic meningitis.

7. Some viruses cause _____ effects, which result in the cell's death.

8. One effect of the viral infection of an animal cell may be the release of enzymes from_____ and autolysis of the cell.

9. One effect of the viral infection of an animal cell may be to transform cells to spindle-shaped cells that have no _____.

10. Many bacteria and viruses can penetrate the intact mucous membranes, such as the organism that causes _____.

11. Contributing to invasiveness by *Streptococcus pyogenes* is a cell wall protein called _____ protein.

12. Also affecting virulence by bacteria are _____ factors that change body temperature.

13. With few exceptions, microorganisms cannot penetrate the unbroken skin; however, some fungi grow on _____ of the skin.

14. The avenue by which a microorganism gains access to the body is called a _____.

15. If the pathogen causes a nonfatal disease, the LD50 test is referred to as _____.

16. The toxin ergot is contained in _____ (resistant mycelia) of a plant pathogen.

17. A product released by macrophages after ingestion of bacterial endotoxins that triggers appearance of fever is called _____.

Key

Matching

I. 1.e 2.a 3.f 4.c 5.c 6.j 7.g
II. 1.a 2.a 3.b 4.a 5.a 6.b
III. 1.e 2.a 3.b
IV. 1.d 2.b 3.e

Fill-in-the-Blanks

1. parenteral   2. animals   3. phagocytosis   4. collagenase   5. hyaluronic acid
6. endo   7. cytocidal   8. lysosomes   9. contact inhibition   10. syphilis   11. M
12. hypothermic   13. keratins   14. portal of entry   15. ID_{50} (infectious dose)
16. sclerotia   17. Interleukin-1

15 *Nonspecific Defenses of the Host*

Pathogenic microorganisms are endowed with special properties that, given the right opportunity, enable them to cause disease. Our bodies, however, have defenses against these microorganisms. In general, our ability to ward off disease through our defenses is called **resistance** and our vulnerability or lack of resistance is known as **susceptibility**. **Non-specific resistance** refers to defenses that tend to protect us from any kind of pathogen. **Specific resistance**, or immunity based on antibody production, is a defense against a particular microorganism.

Skin and Mucous Membranes

Mechanical Factors

The intact skin consists of the **dermis**, an inner portion composed of connective tissue, and the **epidermis**, an outer portion consisting of several layers of epithelial cells arranged in continuous sheets. The top layer of epidermal cells contains the protein keratin. Intact, these barriers are seldom penetrated by microorganisms. **Mucous membranes** line the body cavities that open to the exterior; these include the digestive, respiratory, urinary, and reproductive tracts. Mucous membranes consist of an **epithelial layer** and an underlying connective tissue layer. Cells in mucous membranes secrete mucus, which prevents the cavities from drying out. Some pathogens such as *Treponema pallidum* (syphilis), *Mycobacterium tuberculosis* (tuberculosis), and *Streptococcus pneumoniae* (pneumonia) grow in mucus; some are able to penetrate the membrane, which is generally less protective than the skin. Another mechanical factor involved in protection is the **lacrimal apparatus**, which manufactures and drains away tears. This apparatus has a cleansing effect on the eye surface. **Saliva**, produced by salivary glands, washes microorganisms from the surfaces of the teeth and mucous membrane of the mouth. Mucus tends to trap microorganisms that enter the respiratory and digestive tracts. Cells of the mucous membrane of the lower respiratory tract contain cilia, which are microscopic hairlike projections that move synchronously (ciliary escalator) and propel inhaled microorganisms trapped in mucus out of the respiratory system. The **flow of urine** tends to remove microorganisms from the urinary system, and **vaginal secretions** tend to move microorganisms out of the female body.

Chemical Factors

Oil (sebaceous) glands of the skin produce **sebum**, which forms a protective film over the skin surface. The unsaturated fatty acids of sebum inhibit the growth of certain pathogens. The low pH (3-5) of the skin is partly due to these secretions. Body odor is due to the decomposition of both sloughed-off skin cells and sebum secretions by commensal bacteria. **Sweat** (sudoriferous) **glands** produce **perspiration**, which flushes microorganisms from the skin surface. Perspiration contains **lysozyme**, an enzyme that breaks down the cell walls of gram-positive bacteria. The enzyme also is found in saliva, tears, nasal secretions, and tissue fluids. **Gastric juice** is a mixture of hydrochloric acid, enzymes, and mucus, with an acidity of pH 1.2-3. These stomach secretions destroy many ingested microorganisms and toxins. Important exceptions are the toxins of botulism and *Staphylococcus aureus*. Resident flora usually compete more successfully for available nutrients and may produce metabolic end products that inhibit colonization by pathogens.

Phagocytosis

Operating within the body is the host's second line of defense; **phagocytosis** is an important element of this. The word phagocytosis derives from the Greek words for "eat" and "cell," and refers to the ingestion of a microorganism or any particulate matter by a cell. **Phagocytes** are blood cells or derivatives of blood cells that perform this function.

Formed Elements of the Blood

Blood fluid is called plasma, and cells and cell fragments of the blood are the formed elements. Most important for the present discussion are the leukocytes or white blood cells. Leukocytes are in the following two categories:

1. **Granulocytes** have granules in the cytoplasm and are differentiated into **neutrophils**, which stain with a mixture of acidic and basic dyes; **basophils**, which stain with basic methylene blue; and **eosinophils**, which stain with the acidic dye eosin (Figure 15-1). Neutrophils also are known as **polymorphonuclear leukocytes** (**PMNs**).

2. **Agranulocytes** do not have granules in their cytoplasm and include the **lymphocytes** (important immunity) and **monocytes**. The latter mature into macrophages.

Neutrophils (which account for 60 to 70 percent of leukocytes) are important in phagocytosis, basophils (0.5 to 1 percent of leukocytes) form heparin and histamines that are a factor in inflammation and allergic responses, and eosinophils perform some phagocytosis and increase in numbers during parasitic infections and allergic reactions.

An increase in the number of white blood cells during infection is called **leukocytosis**. Diseases causing this increase include meningitis, mononucleosis, appendicitis, pneumonia, and gonorrhea. Other diseases such as typhoid fever cause decreased leukocyte counts called **leukopenia**. **Differential counts** of blood determine the percentages of each type of white blood cells.

Kinds of Phagocytic Cells

There are two broad categories of phagocytic cells: **granulocytes** (sometimes called **microphages**) and **macrophages**. Cells of both of these categories migrate to sites of infection. Granulocytes are

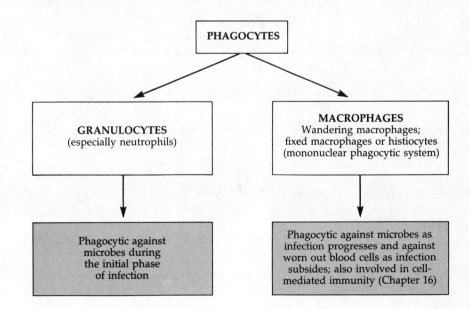

Figure 15.1
Classification of phagocytes.

mostly neutrophils that wander in the blood and can pass through capillary walls to reach trauma sites. Monocytes enlarge and develop into macrophages, which are highly phagocytic cells that are called **wandering macrophages** because of their ability to migrate. **Fixed macrophages** enter tissues and organs and remain there. The fixed macrophages are referred to as the **mononuclear phagocytic** (*reticuloendothelial*) **system** (**RES**). Microorganisms are removed as blood and lymph pass through the RES. Granulocytes predominate early in infections, as indicated by differential counts; as infections subside, monocytes predominate. Figure 15-1 shows the classification of phagocytes.

Mechanisms of Phagocytosis

Chemotaxis and adherence. **Adherence**, or attachment between the cell membrane of the phagocyte and the organism, is facilitated by **chemotaxis**, which is the attraction of microorganisms to chemicals. Capsules help organisms avoid phagocytosis. **Opsonization**, coating a microorganism with plasma proteins such as antibodies and complement, promotes phagocytosis.

Ingestion and digestion. Following adherence, projections of the cell membrane of the phagocyte (**pseudopods**) engulf the microorganism and then fold inward, forming a sac around it called a **phagosome** or **phagocytic vacuole**. This process is called **ingestion**. Within the cell, enzyme-containing phagosomes and lysosomes of the cell fuse to form a larger digestive vacuole, the **phagolysosome** or digestive vacuole, within which the bacteria are usually quickly killed (Figure 15-2). Enzymes in lysosomes include lysozyme, various hydrolytic enzymes, and myeloperoxidases. Some microorganisms, such as the agents of tularemia, brucellosis, and tuberculosis, may remain dormant in phagocytes or even grow actively.

Inflammation

Inflammation is a host response to tissue damage, characterized by redness, pain, heat, swelling, and perhaps loss of function. It is generally beneficial, serving to destroy the injurious agent, to confine or wall it off, and to repair or replace damaged tissue.

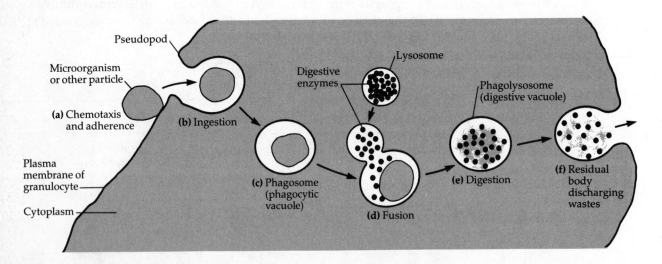

Figure 15.2 Phagocytosis. The drawing shows the mechanism within a granulocyte. (**a**) Chemotaxis and adherence. (**b**) Ingestion. (**c**) Formation of the phagosome. (**d**) Fusion of the phagosome with a lysosome forms a phagolysosome. (**e**) Digestion—destruction of the ingested microorganism. (**f**) Residual body containing indigestible material.

Vasodilation and Increased Permeability of Blood Vessels

Vasodilation is the first stage of inflammation; it involves an increase in blood vessel diameters in the injured area. Increased **permeability** allows defense substances in the blood to pass through the walls of the blood vessels. Vasodilation also is responsible for the redness, heat, swelling, and pain of inflammation. **Histamines** are released by injury and increase permeability as well. **Kinins** cause vasodilation and also increase permeability and attract phagocytes. **Prostaglandins** and **leukotrienes** cause vasodilation. **Clotting elements** in the blood help prevent the spread of the infection; a focus of infection walled off in this fashion is an **abscess**.

Phagocyte Migration

Within an hour of the beginning of inflammation, phagocytes appear at the site. Blood flow decreases as the phagocytes stick to the inner lining of blood vessels (**margination**) and then pass through the vessel wall to the damaged area. This migration is called **diapedesis**. Monocytes appear later in the inflammatory response. Macrophages appear later, after granulocytes have accomplished their function, and ingest dead tissue and invade microorganisms. Collections of dead phagocytic cells and fluids are called **pus**.

Repair

The final stage of inflammation is **tissue repair**. This involves production of new cells by the *stroma* (connective tissue) and the *parenchyma* (functioning part of tissue). Scar tissue results from stroma-type repair.

Fever

The hypothalmus controls body temperature (normally 37°C or 98.6°F). The setting can be altered by ingestion of gram-negative bacteria by phagocytes. The release of endotoxins causes release of interleukin-1 (endogenous pyrogens). **Crisis** refers to a very rapid fall in temperature. Fevers often are beneficial, for they aid body tissue repair and inhibit microbial growth.

Antimicrobial Substances

Complement and Properdin

Complement (abbreviated C) is a group of at least 20 proteins found in normal blood serum. Complement is so named because it complements immune reactions involving antibodies. The properdin system is composed of properdin, factor B, and factor D—all of which are proteins in serum. The properdin system is a component of the alternate pathway described below.

Pathways of complement activation. Complement and properdin act in a sequence called **cascade**; one protein in the cascade activates another. The classical pathway is initiated when an antibody binds to an antigen (Chapter 16). Complement protein C1 then binds to two adjacent antibodies and initiates a sequence known as the **membrane attack complex**. Circular lesions called **trans-membrane channels** are formed and cause the eventual lysis of the cell to which the antibodies attached.

    Inflammation also can develop from complement. Other C-proteins combine with mast cells, and attached antibodies etc., and cause release of histamine, which increases blood vessel permeability. Some C-proteins attract phagocytes. C-proteins also may coat a microorganism and promote its attachment by a phagocyte (opsonization).

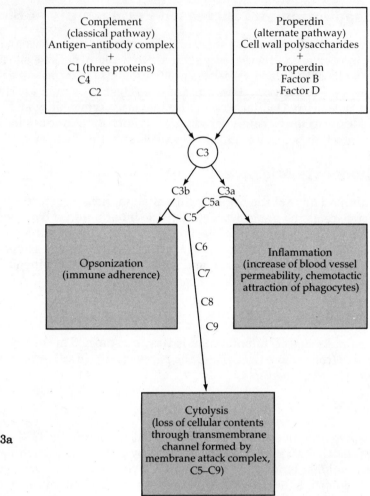

Figure 15-3 Complement activation via classical and alternative (properdin) pathways. Note the cleavage of C3 into C3a and C3b. These fragments induce three kinds of consequences destructive to microorganisms—opsonization, cytolysis, and inflammation.

The **alternate pathway** does not involve antibodies; it is initiated by the interaction between certain polysaccharides and the proteins of properdin system. This is especially important in reactions against gram-negative bacteria. Their cell walls contain the lipopolysaccharide endotoxins that trigger the alternate pathways (Figure 15-3).

Interferon

Interferons are proteins produced by virus-infected animal cells. They tend to interfere with viral multiplication, and are most important in protection against acute virus-caused infections such as influenza. Interferons can be produced by biotechnological methods and have been tested to determine their antiviral and anticancer effects.

Test

In the matching section, there is only one answer to each question; however, the lettered options (a, b, c, etc.) may be used more than once or, perhaps, not at all.

I. Match

1. Produces tears.

2. The outer layer of the skin.

3. An oily substance forming a protective film over the skin surface.

4. Secreted by cells in mucous membrane; prevents the cavities from drying.

5. Inner portion of the skin, composed of connective tissue.

a. Dermis
b. Epidermis
c. Mucus
d. Mucous membrane
e. Lacrimal apparatus
f. Sebum

II. Match

1. The blood fluid.

2. Cells and cell fragments of the blood.

3. Immunity based on antibodies.

4. Movement by a microorganism toward an attractant chemical.

5. Increase in diameter of blood vessels.

6. Collection of dead phagocytic cells and fluids.

7. Vulnerability to a pathogen.

a. Serum
b. Plasma
c. Formed elements
d. Susceptibility
e. Specific resistance
f. Chemotaxis
g. Vasodilation
h. Pus
i. Opsonization

III. Match

1. Neutrophils.

2. Monocytes.

3. Lymphocytes.

4. Microphages.

a. Agranulocytes

b. Granulocytes

IV. Match

1. Increase in the number of white blood cells.

2. Projections of the cell membrane of the phagocyte.

3. A larger digestive vacuole formed when lysosome and phagosome fuse.

4. Decrease in the number of white blood cells.

a. Pseudopods
b. Phagolysosome
c. Phagosome
d. Lysozyme
e. Lysosome
f. Leukocytosis
g. Leukopenia

V. Fill-in-the-Blanks

1. Cells of the mucous membrane of the lower respiratory tract contain _____, which are microscopic hairlike projections.

2. Sudoriferous, or _____ glands, produce perspiration.

3. _____ are also known as polymorphonuclear leukocytes.

4. The most numerous of the granulocytes are the _____.

5. Complement and properdin act in a sequence called a _____.

6. Percentages of the various types of white blood cells are determined by doing a _____.

7. In the membrane attack complex associated with complement, circular lesions called _____ channels are formed.

8. Fixed macrophages are referred to as the _____ system.

9. Phagocytes may stick to the inner lining of blood vessels before passing through the vessel wall to the damaged area; this action is called _____.

10. The _____ controls body temperature.

11. _____ is a group of at least 20 proteins found in normal blood serum.

12. Granulocytes that stain with basic methylene blue dyes are known as _____.

13. Agranulocytes, which do not have granules in their cytoplasm, include the lymphocytes and the _____.

14. The attachment between the cell membrane of the phagocyte and the organism is called _____.

15. The coating of a microorganism with plasma proteins such as antibodies and complement is called _____ and promotes phagocytosis.

16. The host response to tissue damage characterized by redness, pain, heat, and swelling is called _____.

17. The migration of phagocytes through the vessel wall to the damaged area is called _____.

18. Compounds that are produced by the body and that cause vasodilation include _____ and _____.

19. Scar tissue results from _____ type repair.

20. The _____ pathway of the complement system does not involve antibodies.

21. Eosinophils stain with the acidic dye _____.

22. Related to the complement system is the _____ system made up of factors B and D.

23. C-proteins react with mast cells and attached antibodies, and cause the release of _____, which increase blood vessel permeability.

Key

Matching

I.   1.e 2.b 3.f 4.c 5.a
II. 1.b 2.c 3.e 4.f 5.g 6.h 7.d
III. 1.b 2.a 3.a 4.b
IV. 1.f 2.a 3.b 4.g

Fill-in-the-Blanks

1. cilia   2. sweat   3. neutrophils   4. neutrophils   5. cascade   6. differential count
7. trans-membrane   8. mononuclear phagocytic (reticuloendothelial)   9. margination
10. hypothalmus   11. complement   12. basophils   13. monocytes   14. adherence
15. opsonization   16. inflammation   17. diapedesis   18. kinins; prostaglandins
19. stroma   20. alternate   21. eosin   22. properdin   23. histamine

16 *Specific Defenses of the Host: The Immune Response*

In addition to the nonspecific host defenses discussed in the last chapter, humans have **innate resistance** to some illnesses. We do not get certain animal diseases; some racial groups, probably because of historical exposure, are comparatively more susceptible to some diseases. In this chapter, we will discuss the specific defenses of the host; that is, the immune system.

Types of Acquired Immunity

Acquired immunity refers to immunity developed in response to **antigens**, which are organisms or substances that provoke the production of special proteins called **antibodies** or of certain **specialized lymphocytes**.

Naturally Acquired Immunity

Naturally acquired active immunity is obtained by natural exposure to antigens; for example, disease organisms. **Subclinical infections** also can confer immunity. **Naturally acquired passive immunity** involves transfer of antibodies formed by the mother to her infant. This may be done by **placental transfer** and renders the infant immune to most of the diseases to which the mother was immune. Breast milk, especially the first secretions called **colostrum**, provide some immunity. These passive immunities last only a few weeks or months.

Artificially Acquired Immunity

Artificially acquired active immunity results from vaccination (immunization) in which vaccines composed of inactivated bacterial toxins (**toxoids**), killed microorganisms, or living but attenuated (weakened) microorganisms are injected. **Artificially acquired passive immunity** involves injection of antibodies formed in the serum, (the fluid remaining when blood has clotted and blood cells and other matter have been removed), of one person or animal. It is injected in another person and provides an immediate but short term immunity. The antibody-rich serum component is **gamma globulin** or **immune serum globulin**.

The Duality of the Immune System

The **humoral immune system** involves antibodies that are dissolved in various body fluids or secretions (humors). This system responds when **B cells** (specialized lymphocytes) are exposed to antigens. Humoral immunity defends mostly against bacteria, bacterial toxins, and viruses circulating in body fluids.

The **cell-mediated immune system** involves specialized lymphocytes called **T cells**. T cells do not secrete antibodies but have antibodylike antigen receptors attached to their surfaces. The cell-mediated immune system is most effective against bacteria and viruses located within phagocytic or other host cells, and against fungi, protozoans, and helminths. It also responds to what it perceives as foreign, i.e. transplanted tissue, and cancer cells.

Antigens and Antibodies

Most **antigens** (sometimes called **immunogens**) are proteins, lipoproteins, glycoproteins, or large polysaccharides. These may be part of microorganisms or antigens such as pollen, egg white, blood cells, or transplanted tissues or organs. Antibodies usually recognize and interact with antigenic determinants on the antigen, rather than an entire antigen. A bacterium or virus may have numerous antigenic determinants. **Haptens** are low-molecular-weight antigens that are not antigenic unless first attached to a carrier molecule. Once antibody against the hapten has formed, the hapten will react independent of its carrier. Penicillin is a good example of a hapten.

The Nature of Antibodies

Antibodies belong to proteins called **immunoglobulins (Ig)**. They are highly specific and react with only one type of antigenic determinant. Each antigen has at least two antigen-binding sites. The **valence** is the number of such sites on the antibody.

Antibody structure. Figure 16-1a shows a typical monomer type antibody. It has four protein chains—two identical **light (L) chains** and two identical **heavy (H) chains**. At the ends of the arms of Y-shaped molecules there are **variable (V) regions** with a structure that accounts for the ability of different antibodies to recognize and bind with different antigens (Figure 16-1b).

The stem of the antibody monomer and lower parts of the Y arms are called **constant (C) regions**. The stem of the Y-shaped monomer is the **Fc (crystallizable fragment) region**. The antibody molecule can attach to a host cell by the Fc region. Complement can bind to the sides of the Fc region.

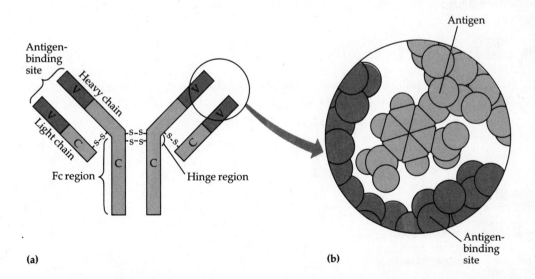

(a)   (b)

Figure 16.1 Structure of a typical antibody molecule. **(a)** The Y-shaped molecule is composed of two light and two heavy chains linked together by disulfide bridges (S - S). Most of the molecule is made up of constant regions (C), which are the same for all antibodies of the same class. The amino acid sequences of the variable regions (V), which form the two antigen-binding sites., differ from molecule to molecule. The stem of the Y, called the Fc region, allows the antibody to bind to the surface of certain host cells. **(b)** One antigen-binding site is shown enlarged and bound to an antigen. The three-dimensional fit between the antigen-binding site and the antigen is similar to the fit between an enzyme and its substrate.

Immunoglobulin Classes (Refer to Figure 16.2)

IgG antibodies account for 80 to 85 percent of all antibodies in serum. They are monomers and readily cross blood vessel walls to enter tissue fluids. Maternal IgG crosses the placenta to confer immunity to the fetus. IgG antibodies protect against circulating bacteria and viruses, neutralize bacterial toxins, trigger the complement system, and bind to antigens to enhance action of phagocytic cells.

IgM antibodies comprise 5 to 10 percent of antibodies in serum. IgM has a pentamer structure of five Y-shaped monomers held together by a **J (joining) chain**. IgM antibodies are the first to appear in response to an antigen but their concentration declines rapidly. IgM antibodies generally do not enter surrounding tissue. IgM molecules are especially effective at cross-linking particulate antigens, causing their aggregation. The fact that IgM appears first in response to a primary infection and is relatively short lived makes it valuable in disease diagnosis. Its presence in high concentrations in a patient makes it likely that the antibodies are associated with the disease pathogen. IgG is so long-lived that its presence may indicate immunity against a disease condition in the more distant past.

IgA antibodies account for about 15% of the antibodies in serum. IgA may be joined by a J chain into dimers of two Y-shaped monomers. IgG is found in body secretions such as saliva and colostrum. A secretory component protects the IgA from enzymes and may help it enter secretory tissues. The main function of IgA is preventing attachment of viruses, etc., to mucosal surfaces.

IgD antibodies are only about 0.1 percent of the total serum antibodies. They are monomers, and are found in blood and lymph cells, and on B cell surfaces. They do not fix complement or cross the placenta. IgD functions are little known but IgD is present on the surface of B cells, indicating they may help initiate the immune response.

IgE antibodies are monomers slightly larger than IgG and constitute only 0.002 percent of the total serum antibodies. They bind by their Fc sites to mast cells and basophils. When an antigen reacts with IgE antibodies the mast cell or basophil releases histamine and other chemical mediators involved in allergic reactions. Persons highly allergic or with gastrointestinal parasites often have high IgE serum levels.

B Cells and Humoral Immunity

B cells, T cells, and macrophages all develop from **stem cells** located in bone marrow of adults or liver of embryos. Each B cell will produce antibodies against one specific antigen.

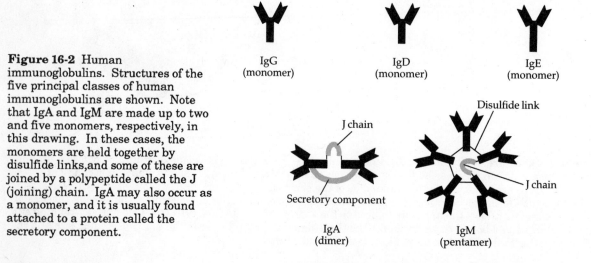

Figure 16-2 Human immunoglobulins. Structures of the five principal classes of human immunoglobulins are shown. Note that IgA and IgM are made up to two and five monomers, respectively, in this drawing. In these cases, the monomers are held together by disulfide links, and some of these are joined by a polypeptide called the J (joining) chain. IgA may also occur as a monomer, and it is usually found attached to a protein called the secretory component.

IgG
(monomer)

IgD
(monomer)

IgE
(monomer)

Disulfide link

J chain

Secretory component

IgA
(dimer)

J chain

IgM
(pentamer)

B Cell and Antigen Interactions

When a B cell contacts an antigen it differentiates into **plasma cells**, which secrete antibodies against the antigen. Antigen stimulation of a B cell also results in memory B cells. **T-dependent antigens** must work cooperatively with certain macrophages and T cells. Examples of T-dependent antigens are bacteria, foreign red blood cells, proteins, and hapten-carrier combinations with many antigenic determinants. The T-dependent antigen (Figure 16-3) is ingested by macrophages or dendritic cells known as **antigen-presenting cells** (APC). Portions of the antigen appear on the APC surface. Helper T cells (Th) interact with APC and are primed to activate an appropriate B cell. **T-independent antigens** can stimulate the response of B cells without the intervention of T cells (Figure 16-4). Most T-independent antigens are repeating polysaccharide or protein subunits such as bacterial flagella or the lipopolysaccharide layer of gram-negative bacteria. Most of the antibodies produced are of the IgM class.

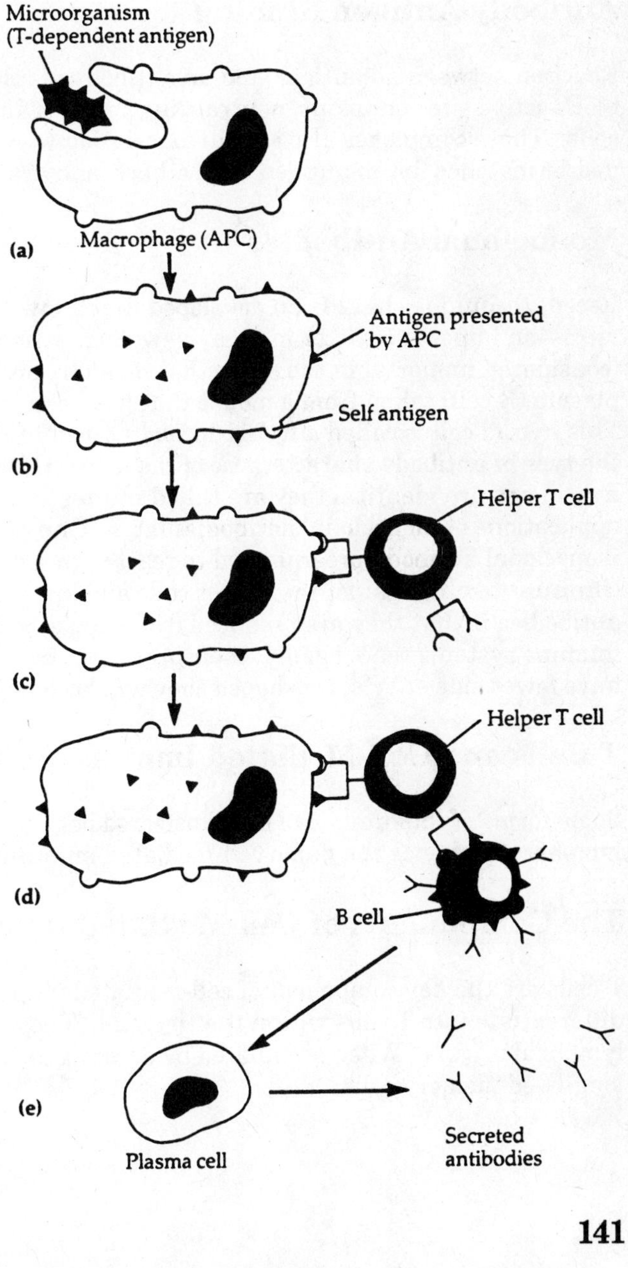

Figure 16-3 How helper T cells may activate B cells to make antibodies against T-dependent antigens. B cells require the help of antigen-presenting cells (APCs) and specialized helper T cells to produce antibodies against T-dependent antigen. The process shown is a likely one, but it is not completely understood. **(a)** The microbial antigen is attached by an APC and partially digested. **(b)** Antigen fragments are presented on the APC surface. Like all body cells, the APC also has self (MHC) antigens on the surface. **(c)** A helper T cell (T$_H$) specific for the presented antigen interacts with both that antigen and the self antigen. **(d)** The helper T cell then activates an appropriate B cell, probably one that itself has both antigens on its surface, as well as antibodies to the microbial antigen. **(e)** The interaction of T and B cells, along with direct stimulation by the microbial antigen, triggers the B cell to differentiate into a plasma cell, which secretes antibodies.

141

T-independent antigen

B cell

Figure 16-4 T-independent antigens. T-independent antigens have repeating units that form multiple bonds with a B cell. These antigens stimulate the B cell to make antibodies without the aid of helper T cells.

Antibody-Antigen Binding

Reaction between an antigen and an antibody results in an antigen-antibody complex. This may block active sites on toxins, neutralizing them, or inactivate viruses by preventing attachment to host cells. These complexes also can, in association with complement, lyse cells. Some antibodies are poorer matches for an antigen than others, and are said to have less **affinity**.

Monoclonal Antibodies

Recent techniques have been developed that allow the production of large volumes of antibodies in vitro—an important factor in many new diagnostic tests. In these techniques, a cancerous cell (considered immortal in the sense that it can be propagated indefinitely) and an antibody-secreting plasma (B cell) taken from a mouse that has been immunized with a particular antigen are fused. This hybrid cell is called a **hybridoma**. Grown in culture, such a hybridoma will continue to produce the type of antibody characteristic of the ancestral B cell. Because the antibodies produced by such a hybridoma are identical they are called **monoclonal antibodies**. We will see a number of applications of monoclonal antibodies later. One possibility is as a treatment for cancer. A monoclonal antibody programmed to react with a cancer can be combined with a toxin (**immunotoxin**) against the cancer cells and kill them. A problem in therapeutic use of monoclonal antibodies is that they are produced from mouse cells. These are considered foreign, and human immune systems react against their presence. Human-derived monoclonal antibodies would probably have fewer side effects; it is hoped they will become commercially available in the near future.

T Cells and Cell-Mediated Immunity

Some forms of immunity can be transferred between animals only by transferring certain lymphocytes, hence the name cell-mediated immunity.

The Components of Cell-Mediated Immunity

T cells are the key component of cell-mediated immunity. After developing from stem cells, they differentiate into T cells within the **thymus**. They then migrate to the lymph nodes, spleen or other lymphoid organs. When stimulated by an antigen they differentiate into effector T cells and some long-lived memory cells.

Types of T Cells

Helper T cells help present T-dependent antigens to B cells. They also help other T cells respond to antigens. Other helper T cells, called **delayed hypersensitivity T cells (Td)** are associated with allergic reactions such as poison ivy and with rejection of transplanted tissues. Td cells produce substances called **lymphokines** that recruit defensive cells such as macrophages. **Suppressor T cells (Ts)** are not completely understood. They inhibit conversion of B cells into plasma cells, which may be a way of turning off the immune response when an antigen no longer is present. They also suppress activity of some other T cells so that immune responses are not developed against host antigens. Both Th and Ts cells are known as **regulatory T cells**. **Cytotoxic T cells (Tc)** destroy target cells upon contact. They are important in action against cancer, rejection of transplanted tissue, and protection against intracellular bacterial and viral infections. Tc cells recognize viral antigens on the host cell and cause destruction of that cell. This is done by release of a protein called **perforin**, that forms a pore in the target cell membrane.

Killer Cells

Other effector cells are **natural killer cells**. These are not immunologically specific and attack a broad group of cells, especially tumor cells. They must contact the target cell to lyse it. **Killer cells (K cells)** are effector cells that attack any target cells coated with antibodies and lyse them by some unknown mechanism.

Lymphokines

When Td cells (and possibly others) are stimulated by an antigen they release proteins called **lymphokines** (also known as **cytokines**). The best known lymphokines are the **interleukins**. Several representative interleukins and their functions are listed in Table 16-4 in the text.

The Cell-Mediated Immune Response

T cells have specificity for only a single antigen, one which recognizes receptors on the T cell. T cells are unable to bind with soluble antigens. They usually respond only to antigens processed by an APC, and also there must be a close association with **major histocompatability antigens (MHC)** on the APC. MHC antigens are unique to each individual and are an expression of "self". This requirement for a close association between the two types of antigens is called **associated recognition**. The advantage is that this minimizes the chances of a T cell mounting an immune attack on the body's own tissues. APC cells stimulated by an antigen secrete **interleukin-1 (IL-1—a monokine)**. IL-1 causes the T cell to synthesize **interleukin-2 (IL-2)**. The IL-2 binds to the T cell, which begins to proliferate and differentiate into different effector cells (see Figure 16-14 in the text).

Clonal Selection

Individual B or T cells are specific for only a single antigen. Contact with this antigen leads to proliferation and differentiation of that B cell or T cell. This selection that results in multiplication of a clone of cells with identical immunological specificity is called **clonal selection**. The specificity of a B cell depends upon antibodies bound to the cell membrane. T cell receptors also share many properties with immunoglobulins. The astonishing diversity of antigenic recognition by B cells and T cells involves recombination of DNA segments within their genomes to create genes for many more kinds of antibodies than would otherwise be possible. The immune system does not usually react with the host's own tissues, a phenomenon called **tolerance**. Apparently the B cells and T cells that interact with self antigens are destroyed during fetal development.

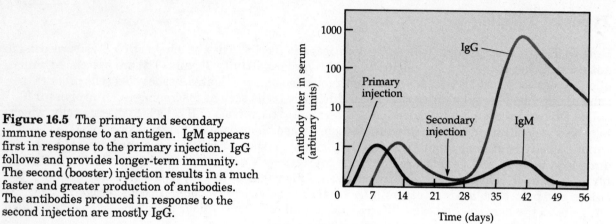

Figure 16.5 The primary and secondary immune response to an antigen. IgM appears first in response to the primary injection. IgG follows and provides longer-term immunity. The second (booster) injection results in a much faster and greater production of antibodies. The antibodies produced in response to the second injection are mostly IgG.

Immunological Memory

Antibody **titer** is the amount of antibody present in the serum. First contact with antigen results in a **primary response** (see Figure 16-5). A second exposure to an antigen results in an intensified response called the **secondary response**, (also called memory or anamnestic response). In cell mediated immunity the memory is mainly in certain effector T cells that distinguish for an extended time between host cells and foreign cells. This is due to long lived memory cells produced as a component of the primary response.

Test

In the matching section, there is only one answer to each question; however, the lettered options (a, b, c, etc.) may be used more than once or, perhaps, not at all.

I. Match

1. Antigen converts these into plasma cells.

2. Involved in cell-mediated immunity.

3. Directed against transplanted tissue cells and cancer cells.

4. Have been influenced by the thymus.

5. Defends mainly against bacteria and viruses circulating in blood and lymph.

6. Responsible for rejection of foreign tissue transplants.

a. B cells

b. T cells

II. Match

1. Based on antibodies produced as a result of recovery from a disease.

2. Passed to a fetus across the placenta.

3. Passed to infant in human colostrum.

4. Passed to recipient by injection of gamma globulin blood fraction from other persons.

5. Based on production of antibodies from injection of a toxoid.

a. Naturally acquired active immunity

b. Artificially acquired passive immunity

c. Naturally acquired passive immunity

d. Artificially acquired active immunity

III. Match

1. An incomplete antigen that will react with antibodies but will not stimulate their formation.

2. The number of determinant sites on an antigen or antibody

3. The source of B cells and T cells.

4. Proteins released by Td cells that are factors in inflammation.

5. Fusion of plasma cell and cancer cell.

a. Hapten

b. Valence

c. Stem cells

d. Lymphokines

e. Hybridomas

f. Memory cells

IV. Match

1. A pentamer; the first antibody class to appear, though comparatively short-lived.

2. Most abundant immunoglobulin in serum.

3. Functions of this immunoglobulin class are not well defined, but may help initiate the immune system.

4. Involved in allergic reactions; also found at elevated serum levels during infections by certain parasites.

5. Often forms dimers of two immunoglobulin monomers.

a. IgA

b. IgG

c. IgD

d. IgE

e. IgM

V. Match

1. Synonym for antigen.

2. Antibodies attach to phagocytic cells and enhance their activity.

3. Protein bound to IgA immunoglobulins.

4. Blood fraction that contains most of the serum immunoglobulins.

5. Antigenic; will stimulate production of antitoxins.

a. Gamma globulin

b. Immunogens

c. Opsonization

d. Toxoid

e. Secretory component

VI. Match

1. Probably inhibit conversion of B cells into plasma cells.

2. Needed to present T-dependent antigens to B cells.

3. Directly destroy target cells such as transplanted tissue.

a. Delayed hypersensitivity cells (Td)
b. Helper T cells (Th)
c. Cytotoxic C cells (Tc)
d. Suppressor T cells (Ts)

VII. Fill-in-the-Blanks

1. Resistance present at birth and not involving humoral or cell mediated immunity is _____ immunity.

2. A _____ site is a specific chemical group on an antigen that combines with the antibody.

3. The region on an immunoglobulin monomer that allows the molecule to attach to a host cell is known by the letters _____.

4. The five monomers comprising the IgM molecule are held together by a _____.

5. The antibody _____ is the measured amount of antibody in the serum.

6. The first breast milk secretions are called _____; they provide some immunity to the newborn.

7. IgA is protected from enzyme activity by a _____ component.

8. The fluid remaining when blood has been clotted, and when blood cells and other matter have been removed is called _____.

9. Effector cells that attack any target cells coated with antibodies and lyse them with some unknown mechanism are_____ cells.

10. When an antigen-presenting cell is stimulated by an antigen, it secretes a monokine called _____.

11. The immune system does not usually react with the host's tissues, a phenomenon called _____.

12. Low-molecular-weight substances such as penicillin that do not (by themselves) cause formation of antibodies are known immunologically as _____.

13. Which cells actually produce humoral antibodies, B cells or plasma cells? _____

14. Are antigens that are made up of repeating polysaccharide or protein subunits more likely to be T-dependent antigens or T-independent antigens? _____

15. The second time we encounter an antigen our response is faster and more intense. This is termed the _____ response.

16. The lymphokine that attracts macrophages to an infection site is the macrophage _____ factor.

17. The lymphokine that activates macrophages to make them more efficient in destroying antigens is the macrophage _____ factor

18. Some antibodies are poorer matches for an antigen than others; they are said to have less _____.

19. A monoclonal antibody programmed to react with a cancer can be combined with a toxin, _____, against the cancer cells and kill them.

20. Cytotoxic T cells cause the death of target cells by release of a protein called _____ that forms a pore in the target cell membrane.

Key

Matching

I.   1.a   2.b   3.b   4.b   5.a   6.b
II. 1.a   2.c   3.c   4.b   5.d
III. 1.a   2.b   3.c   4.d   5.e
IV. 1.e   2.b   3.c   4.d   5.a
V.   1.b   2.c   3.e   4.a   5.d
VI. 1.d   2.b   3.c

Fill-in-the-Blanks

1. innate   2. antigenic determinant   3. Fc   4. J-chain   5. titer   6. colostrum
7. secretory   8. serum   9. killer (K)   10. interleukin-1   11. tolerance
12. haptens   13. plasma   14. T-independent   15. anamnestic   16. chemotactic
17. activation   18. affinity   19. immunotoxin   20. perforin

17 *Practical Applications of Immunology*

Vaccines

Vaccines had their origin in the practice of **variolation**, in which material from smallpox scabs was inoculated into the bloodstream to provide immunity to the disease.

A disease can be controlled if most, but not all, of the population is immune. This is called **herd immunity**.

Characteristics of Vaccines

A **vaccine** is a suspension of microorganisms, or some part or product of them, that will induce immunity when it is injected into the host. The microorganisms may be **inactivated** (killed) or **attenuated** (weakened). Attenuated vaccines tend to mimic actual infection and usually provide better immunity. **Toxoids** are inactivated bacterial toxins.

New Vaccine Development

The thought behind many of the newer approaches to vaccines is to avoid using the entire organism as an antigen and thereby improve safety and minimize side effects.

Subunit Vaccines

A **subunit vaccine** uses only those antigenic fragments of a microorganism that are best suited to stimulate an immune response. These subunits may be produced by bacteria or yeasts, for example, by use of genetic-engineering techniques.

Acellular Vaccines

A conventional vaccine can be fragmented and the desired antigens separated out, rather than using the complete cell. These are called **acellular vaccines**.

Anti-Idiotypic Vaccines

The basic idea of an **anti-idiotypic** vaccine is that, rather than using an antigen as the vaccine, an antibody is used that mimics the shape of the antigen (and thus induces immunity) but is itself harmless. The technique is illustrated in Figure 17-1 in the text.

Diagnostic Immunology

The techniques we use to detect the presence and amounts of antibodies in blood serum are called **serology**. We cannot see antibodies and must infer their presence indirectly by a variety of reactions.

Precipitation Reactions

Precipitation reactions involve the reaction of **soluble** antigens with IgG or IgM antibodies to form large interlocking aggregates called **lattices**. Because the lattices precipitate from solution, the antibodies involved have been called **precipitins**. The **optimal ratio** required to cause precipitation can be easily arranged by allowing the antigen and antibody to diffuse toward each other. In a capillary tube this is called a **ring test**. In a petri dish the antigens and antibodies can be placed into wells cut into the agar medium. This is the basis of the **Ouchterlony test** (see Figure 17-4 in the text). If passive diffusion of the reagents is too slow, the test can be speeded up by applying an electric current, called **immuno-electrophoresis**. This is used in research in separation of proteins in human serum. The **countercurrent immuno-electrophoresis test** (**CIE**) is used in diagnosis of bacterial meningitis. The pH is adjusted so the antigens and antibodies have opposite charges and an electrical current causes them to move toward each other rapidly.

Agglutination Reactions

Agglutination reactions involve particulate antigens that can be linked together by antibodies in a process called agglutination.

Direct Agglutination Tests

Direct agglutination tests detect antibodies against relatively large **cellular** antigens such as red blood cells, bacteria, and fungi. The amount of antigen in each well of a microtiter plate is the same. The serum containing the antibodies is sequentially diluted out in a series of wells. The higher the concentration of antibodies in the serum, the more dilutions are required to dilute it to the point where no reaction occurs with the antigen. This is a measure of **titer**, or concentration of antibody. For diagnostic purposes a **rise in titer** during the course of a disease is very significant.

Indirect (Passive) Agglutination Tests

Soluble antigens can respond to agglutination tests if the antigens are adsorbed onto particles, especially minute latex spheres. In such **indirect**, **passive**, **agglutination tests** the antibody reacts with the soluble antigen adhering to the particle. This is the basis of many new diagnostic tests.

Hemagglutination

Certain viruses have the ability to agglutinate red blood cells; a process called **viral hemagglutination**. This is not an antibody-antigen reaction. If a person's serum contains antibodies against such viruses, these antibodies will react with the viruses and neutralize them. This is the basis of the **hemagglutination-inhibition test** used in the diagnosis of influenza, measles, mumps, etc.

Complement Fixation Reactions

Complement (Chapter 15) is a group of serum proteins. During most antigen-antibody reactions, the complement binds to the antigen-antibody complex and is fixed. This complement fixation can be used to detect antibody. Antibodies that do not produce a visible reaction by other means can sometimes be demonstrated by the fixing of complement during the antigen-antibody reaction (see Figure 17-9 in the text).

Neutralization Reactions

The harmful effects of a bacterial exotoxin or a virus can be eliminated by specific antibodies against it. Such a substance is called an **antitoxin**, as used in treatment of diphtheria. Antitoxins produced in other animals can be injected to provide passive immunity against a toxin, for example. If an antibody binds to a virus it prevents it from attaching to a host cell and infecting it. If a virus causes cytopathic effects in a cell culture, a neutralizing viral antibody can be used to detect it—lack of cytopathic effects indicates the presence of antibody in such a **neutralization test**.

Immunofluorescence and Fluorescent-Antibody Techniques

Fluorescent-antibody (FA) techniques use fluorescent dyes that are combined with antibodies to make them fluoresce when exposed to ultraviolet light. This requires a fluorescence microscope with ultraviolet light illumination. In **Direct FA tests** the antigen to be identified is fixed onto a slide. The fluorescein-labeled antibodies are added. Unbound antigen is washed free and the slide is examined under the ultraviolet microscope. In the **Indirect FA test** a known antigen is fixed onto a slide. A test serum with an unknown antibody is added. If the serum contains antibodies against the test antigen it will bind to it, but this is not visible. So the antigen-antibody complex can be seen, fluorescein-labeled anti-human gamma globulin, an antibody that reacts specifically with any human antibody is added to the slide. Unbound reagents are washed from the slide, which is examined under the fluorescence microscope. If the known antigen on the slide appears fluorescent, antibody specific to it was in the test serum.

Enzyme-Linked Immunosorbent Assay (ELISA)

One of the most widely used serological tests is the **enzyme-linked immunosorbent assay (ELISA)**, also known as the **enzyme immunoassay (EIA)**. A microtiter plate with numerous shallow wells is used in the two basic methods we will describe. Variations of the test may use reagents bound to tiny latex particles rather than to the surfaces of the microtiter plates.

In the **Double Antibody Sandwich Method**, the object is to identify an unknown antigen, such as a drug in a serum sample. In the **Indirect Immunosorbent Assay** the object is to determine the presence of certain antibodies in the serum. This assay currently is used to test for antibodies against the AIDS virus. In both of these variations of the ELISA test, a positive reaction is indicated by a color change in the well due to action of an enzyme linked to the last antibody added, and allowed to react with a substrate added in the final step. Figure 17-1 shows the mechanism of the ELISA tests.

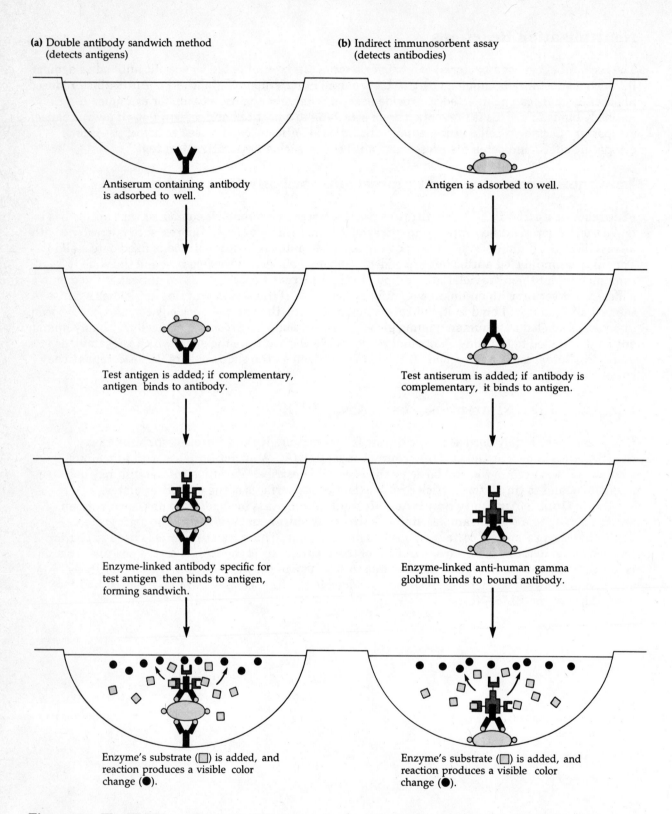

(a) Double antibody sandwich method
(detects antigens)

(b) Indirect immunosorbent assay
(detects antibodies)

Antiserum containing antibody
is adsorbed to well.

Antigen is adsorbed to well.

Test antigen is added; if complementary,
antigen binds to antibody.

Test antiserum is added; if antibody is
complementary, it binds to antigen.

Enzyme-linked antibody specific for
test antigen then binds to antigen,
forming sandwich.

Enzyme-linked anti-human gamma
globulin binds to bound antibody.

Enzyme's substrate (▢) is added, and
reaction produces a visible color
change (●).

Enzyme's substrate (▢) is added, and
reaction produces a visible color
change (●).

Figure 17-1 The ELISA, or EIA, test. The reagents are usually contained in small wells of a microtiter plate.
(a) The double antibody sandwich method is used to detect antigens. **(b)** The indirect immunosorbent assay
detects antibodies.

152

Test

In the matching section, there is only one answer to each question; however, the lettered options (a, b, c, etc.) may be used more than once or, perhaps, not at all.

I. Match

1. Makes use of fact that certain viruses will cause agglutination of red blood cells.

2. The pH is adjusted so that the antigens and antibodies have opposite charges, then an electric current is applied to the solution.

3. The absence of complement is indicated by hemolysis.

4. A precipitation type test in which wells are cut into the agar on a petri dish.

5. Soluble antigens are detected by binding them to small latex particles, for example, and causing their agglutination.

6. The ELISA test used to screen for AIDS antibodies in serum.

a. Ouchterlony test

b. Countercurrent immunoelectrophoresis test

c. Indirect agglutination test

d. Complement fixation test

e. Double antibody sandwich ELISA test

f. Indirect immunosorbent assay ELISA test

g. Hemagglutination-inhibition test

II. Match

1. An antibody that mimics the shape of, and acts as, an antigen.

2. Unwanted components are removed from a whole cell vaccine.

3. An inactivated toxin.

a. Subunit vaccine
b. Toxoid
c. Acellular vaccine
d. Anti-idiotypic-vaccine
e. Antitoxin

III. Fill-in-the-Blanks

1. Before the invention of modern vaccines, material from smallpox scabs was inoculated into the bloodstream to give immunity to the disease; this was called _____.

2. The measure of the concentration of antibody in serum is called _____.

3. Fluorescein labeled anti-human gamma globulin would be used in the _____ fluorescent antibody test.

4. Antibodies against viruses may _____ the viruses' cytopathic effects in cell cultures and serve to identify them.

5. For diagnostic purposes, a rise in _____ during the course of a disease is very significant.

6. A disease can be controlled if most, but not all, of the population is immune; this is called _____ immunity.

7. Techniques to detect the presence and amounts of antibodies in blood serum are called _____.

Key

Matching

I. 1.g 2.b 3.d 4.a 5.c 6.f
II. 1.d 2.c 3.b

Fill-in-the-Blanks

1. variolation   2. titer   3. indirect   4. neutralize   5. titer   6. herd   7. serology

18 *Disorders Associated with the Immune System*

Hypersensitivity

The term **hypersensitivity (allergy)** refers to sensitivity beyond what is considered normal. It occurs in persons who have been previously sensitized by exposure to an antigen, called in this context an **allergen**. Once sensitized, another exposure to the antigen triggers an immune response that damages host tissue. There are four principal types of hypersensitivity reactions.

Type I (Anaphylaxis) Reactions

Anaphylaxis is from the Greek words for protection-against. These reactions result from combining antigens with IgE antibodies; they may be **systemic**, producing sometimes fatal shock and breathing difficulties; or **localized**, such as hayfever, asthma, or hives.

IgE antibodies are cytotrophic, binding to the surfaces of mast cells and basophils. **Mast cells** are prevalent in connective tissue of skin, respiratory tract, and surrounding blood vessels. **Basophils** circulate in the blood. When an antigen combines with antigen-combining sites on two adjacent IgE antibodies and bridges the space between them, the mast cell or basophil undergoes degranulation, releasing chemicals called mediators. The best known mediator is **histamine**, which affects the blood vessels, causing edema (swelling), erythema (redness), increased mucus secretion, and smooth-muscle contractions resulting in breathing difficulty. Other mediators are **leukotrienes** (which tend to cause contractions such as the spasms of asthmatic attacks), and **prostaglandins** (which tend to cause increased secretions of mucus). Collectively, mediators attract neutrophils and eosinophils to the site and cause inflammatory symptoms.

When an individual sensitized to an injected antigen, such as an insect sting or penicillin, receives a subsequent injection, the release of mediators can result in a drop in blood pressure (shock) that can be fatal in a few minutes. This is termed **systemic anaphylaxis**, or **anaphylactic shock**.

Localized Anaphylaxis

Localized anaphylaxis usually is associated with antigens that are ingested or inhaled, rather than injected. Examples are hay fever, for which **antihistamine** drugs often are useful to treat symptoms, and asthma, for which **epinephrine (adrenalin)** is usually administered. It is sometimes difficult to distinguish between food hypersensitivity and food intolerance. Hives are a characteristic of true food allergy.

Prevention of Anaphylactic Reactions

If contact with the allergen cannot be avoided, **desensitization** might be attempted. This consists of injections of a series of small doses of the antigen. The idea is to induce IgG anti-bodies that serve as blocking antibodies that intercept and neutralize antigens before they can react with cell bound IgE.

Type II (Cytotoxic) Reactions

Immunological injury resulting from type II reactions is caused by antibodies that are directed at antigens on the host's blood cells or tissue cells. The host cell plasma membrane may be damaged by antibody and complement; or phagocytic cells or K cells are attracted to the Fc regions of antibodies attached to the host cell. These cells then may discharge lytic enzymes to kill the host cell. Transfusion reactions, such as those involving the ABO and Rh blood group systems are of this type.

The ABO Blood Group System

Human blood is grouped into four principal types: A, B, AB, and O — a classification called the **ABO blood group system**. Persons with type A, for example, have antigens designated A on their red blood cells. Persons with blood type O lack both A and B surface antigens. The main features of the ABO blood group system are summarized in Table 18-1. In about 80 percent of the population, called **secretors**, antigens of the ABO type appear in saliva, semen, and other bodily fluids.

Hemolytic Disease of the Newborn

Roughly 85 percent of the human population has an antigen named the **Rh factor** and are called Rh⁺. If blood from an Rh⁺ donor is given to an Rh⁻ recipient, the production of anti-Rh antibodies will be stimulated. If the Rh⁻ person receives a transfusion of Rh⁺ blood, there will be a reaction (see Figure 18-1).

If an Rh⁻ female and an Rh⁺ male produce a child, the child may be Rh⁺ and sensitize the mother at birth. If the fetus in a later pregnancy is Rh⁺, the antibodies developed in the mother may attack the fetal red blood cells. This results in **hemolytic disease of the newborn**, once called erythroblastosis fetalis. This may be prevented by passive immunization with anti-Rh antibodies or even replacement transfusion of the fetus' blood.

Other Cytotoxic Diseases

Idiopathic thrombocytopenic purpura occurs when blood platelets, which are essential for clotting, are coated with drug molecules that function as haptens. If antibodies developed against these haptens cause destruction of the platelets, a disease condition may result. In **autoimmune hemolytic anemia** the body may form antibodies against its own red blood cells. **Myasthenia gravis** is a disease in which antibodies coat the acetylcholine receptor junctions preventing nerve impulses from reaching the muscles. In **Graves' disease**, antibodies attach to receptors on the thyroid gland and cause excessive production of thyroid-stimulating hormones.

Table 18-1 The ABO Blood Group System

Characteristic	Blood Type			
	A	B	AB	O
Antigen present on the red blood cells	A	B	Both A and B	Neither A nor B
Antibody normally present in the plasma	anti-B	anti-A	Neither anti-A nor anti-B	Both anti-A and anti-B
Plasma causes agglutination of red blood cells of these types	B, AB	A, AB	None	A,B, AB
Percent in a mixed Caucasian population	41	10	4	45
Percent in a mixed black population	27	20	7	46

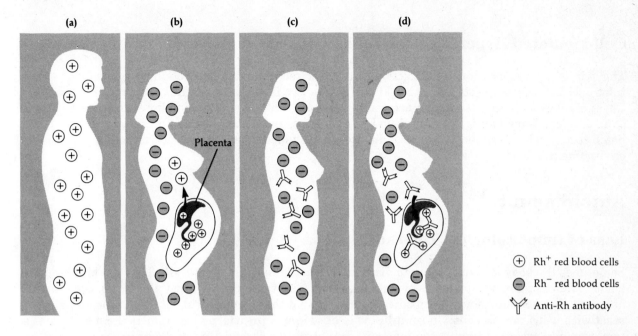

Figure 18-1 Hemolytic disease of the newborn. **(a)** Rh⁺ father. **(b)** Rh⁻ mother carrying her first Rh⁺ fetus. Rh antigens from the developing fetus can enter the mother's blood across the placenta during delivery. **(c)** In response to the fetal Rh antigens, the mother will produce anti-Rh antibodies. **(d)** If the woman becomes pregnant again with an Rh⁺ fetus, her anti-Rh antibodies will pass through the placenta into the blood of the fetus. The result is damage to fetal red blood cells.

Legend:
⊕ Rh⁺ red blood cells
⊖ Rh⁻ red blood cells
Y Anti-Rh antibody

Type III (Immune Complex) Reactions

Immune complexes form when certain ratios of antigen and antibody occur. When a certain antigen-antibody ratio exists, usually with a slight excess of antigen, the soluble complexes that form are small and escape phagocytosis. Circulating in the blood, they may locate in the basement membrane beneath endothelial cells of blood vessels. This can set up an inflammatory, tissue damaging reaction. Diseases of this type are **glomerulonephritis** (affecting kidney glomeruli), **systemic lupus erythematosus**, and **rheumatoid arthritis**. The **Arthus reaction** is an immune complex reaction in which a local area of necrotic tissue surrounds the site at which antigen is injected into a person already sensitized to that antigen.

Type IV (Cell-Mediated) Reactions

Up to this point, our discussion of hypersensitivity has involved IgE, IgG, or IgM. Type IV reactions involve cell-mediated immune responses caused mainly by T cells, but sometimes by macrophages. These reactions are often not apparent for a day or more, time during which participating cells migrate to and accumulate near the foreign antigens.

Causes of Type IV Reactions

Usually in type IV reactions the foreign antigens are phagocytized by macrophages and then presented to receptors on the T cell surface. The T cells are primarily Td cells, but may include T cells. A principal factor is the release of lymphokines by T cells reacting with the target antigen.

Cell-Mediated Hypersensitivity Reactions of the Skin

The skin test for tuberculosis is a reaction by a sensitized individual to protein components of tuberculosis bacteria injected into the skin. A day or two is required for the reaction to appear. Cases of allergic contact dermatitis are usually caused by haptens that combine with proteins in the skin. Typical foreign antigens are poison ivy, cosmetics, and metals such as nickel in jewelry. The patch test, in which samples of suspected material are taped to skin, may determine the offending environmental factor.

Autoimmunity

Loss of Immunological Tolerance

Sometimes the body loses its ability to discriminate between self and nonself. This loss of tolerance leads to autoimmunity, a response by antibodies or sensitized T cells against a person's own tissue antigens. In rheumatic fever, antibodies formed against the streptococcal bacteria coincidentally react with heart muscle cells. In **lymphocytic choriomeningitis**, a virus infects the membranes of the brain and provokes an immune reaction. **Hashimoto's thyroiditis** and **Addison's disease** result from T cells attacking the thyroid and the adrenal glands, respectively. **Multiple sclerosis** involves an immunological attack on the myelin coating of nerve cells. In many autoimmune-type diseases there is an association with possession of certain major histocompatibility complex determinants.

Major Histocompatibility Complex

One inherited genetic characteristic is differences in histocompatibility antigens on cell surfaces. The genes controlling these antigens are the **major histocompatibility antigens (MHC)**, in humans also called **human leukocyte antigen (HLA)** complex. For successful transplant surgery tissue, typing is used to match donor and recipient. Matching for class I antigens (HLA-A, -B and -C) has long been standard procedure, but matching for class II antigens (HLA-D and -DR) might be more important. The donor and recipient must be of the same ABO blood type.

Transplantation

The cornea and brain are examples of **privileged sites**—antibodies do not circulate to this region. **Privileged tissue** such as pig heart valves are not antigenic, and do not stimulate an immune response. The transfer of tissue such as skin from one part of an individual to another on the same individual is an **autograft**. **Isografts** are transplants between identical twins. Such transplants are not rejected. **Allografts**, transplants between related persons represent most transplants. **Xenografts** are transplants of tissue from animals other than humans, which tend to be strongly rejected. When bone marrow is transplanted to persons, the transplanted tissue may carry cells capable of mounting an immune response. This is called **graft versus host disease**.

Immunosuppression

Persons receiving transplants require suppression of their immune system (**immunosuppression**) to prevent rejection of the new tissue. The drug cyclosporine suppresses cell-mediated immunity, but at the cost of some liver and kidney toxicity. Another approach is use of muromonab-CD3 (Orthoclone OKT3), which is a monoclonal antibody that reacts with the CD3 surface molecule, which T cells must have to respond to an antigen.

Natural Immune Deficiencies

Occasionally, people are born with defective imune systems. Also, Hodgkin's disease, a form of cancer, lowers cell-mediated immunity, as does removal of the spleen.

AIDS

Acquired Immune Deficiency Syndrome (AIDS) is a type of immunodeficiency disease. A virus destroys helper T cells. The loss of an effective immune system leaves the victim susceptible to diseases that the immune system would normally prevent.

Origin of AIDS

A retrovirus known as human immunodeficiency virus (**HIV**) is the cause of AIDS.

Symptoms of AIDS

The HIV kills helper T cells known as T4 cells, from the CD4 antigen on their surface. It also is able to enter macrophages, B cells endothelial cells that line blood vessesl and certain brain cells. It can remain latent for a long time without necessarily killing them. Macrophages may serve as vehicles of transmission. Within a few weeks of initial infection there are only mild symptoms and the blood test undergoes seroconversion after a short time. After the initial stages, persons may exhibit **AIDS-related complex**, which includes swollen lymph nodes, loss of appetite and weight, fever, rashes, night sweats and fatigue. The majority of those infected will probably develop AIDS and then die within 200 to 400 days. Death usually is from opportunistic infections such as pneumocystis pneumonia or cancers such as Kaposi's sarcoma.

Modes of Transmission

The AIDS virus usually requires exchange of bodily fluids such as blood, semen, or vaginal secretions for transmission. This includes intimate sexual contact, contaminated needles, blood transfusions, and transmission to the fetus by the mother. It is not transmitted by casual social contact.

Vaccines and Treatments

An AIDS vaccine will have to work against the virus-infected cell because there is relatively little free virus in circulation. The virus occurs in more than one antigenic form. The one presently circulating in the United States is HIV-1; HIV-2 is prevalent in Africa. A successful vaccine is not expected until the 1990s, if at all. The search for effective drugs is being pressed.

Immune Response to Cancer

Immunological Surveillance

The immune system's patrolling the body for cancer cells is called **immunological surveillance**. Individual cancer cells, before they become established, are recognized as foreign and destroyed by an effective immune system.

Immunological Escape

Once established, the cancer becomes resistant to immune rejection. This is called **immunological escape**. One possible mechanism is **antigen modulation**, in which tumor cells shed their associated antigens.

Immunological enhancement is illustrated by an experiment in which tumor cells are injected into animals that have either been immunized by these tumor cells or not immunized. The nonimmunized animal tends to reject the tumor cell, the immunized animal succumbs to the tumor cell. A possible explanation is that the tumor surface antigens are coated with antibodies; Tc cells fail to recognize the tumor as nonself and fail to destroy it.

Immunotherapy

A future approach to cancer treatment is immunotherapy, such as the use of monoclonal antibodies with immunotoxins, which we have discussed previously.

Test

In the matching exercises there is only one answer to a question; however, the lettered options (a, b, c, etc.) may be used more than once or, perhaps, not at all.

I. Match

1. Hypersensitivity.

2. Hypersensitivity specifically involving interaction of humoral antibodies of the IgE. class with mast cells.

3. Skin graft from a brother to a sister.

4. Heart of a baboon transplanted to a human.

5. A term used for an antigen causing hyper-sensitivity reactions.

6. Skin graft transferred from the thigh to the nose of the same person.

a. Allergen
b. Anaphylaxis
c. Xenograft
d. Allergy
e. Autograft
f. Allograft
g. Autoimmunity
h. Degranulation

II. Match

1. Drug used for transplantation surgery.

2. Drug that suppresses cell-mediated immunity.

3. Reason why transplantation of a cornea usually is successful.

4. Mediator of Type I reaction that affects blood and results in swelling and reddening.

5. Development of blocking antibodies by repeated exposure to small doses of the antigen.

a. Histamine
b. Leukotrienes
c. Prostaglandins
d. Cyclosporine
e. Privileged site
f. Privileged tissue
g. Desensitization

III. Match

1. The naturally learned ability of the body not to respond immunologically against its own antigens.

2. Ability of some cancer cells to avoid destruction by the immune system.

3. Inhibition of the immune response by drugs, radiation, etc.

4. Treatment of cancer or other disease conditions by using monoclonal antibodies with which toxic compounds have been combined.

5. Phenomenon in which previous exposure of an animal to cells of a tumor make it more likely that the tumor will not be eliminated by the immune system if the tumor is reintroduced later.

a. Immunological escape

b. Immunological surveillance

c. Immunosuppression

d. Immunological tolerance

e. Immunotherapy

f. Immunological enhancement

IV. Match

1. Descriptive of antibodies that tend to attach to mast cells.

2. Epinephrine.

3. Release of mediators from mast cells or basophils during an anaphylactic reaction.

4. Destruction of Rh+ red blood cells by antibodies of maternal origin in a newborn infant; the antibodies are derived from the mother.

5. Individuals in whom ABO antigens are present in body fluids such as saliva and semen.

a. Cytotrophic

b. Erythroblastosis fetalis

c. Degranulation

d. Adrenalin

e. Secretors

V. Match

1. Tuberculin test.

2. Asthma.

3. Glomerulonephritis.

4. Poison ivy dermatitis

5. Graves' disease.

6. Reaction to insect sting.

a. Type I (anaphylaxis) reaction

b. Type II (cytotoxic) reaction

c. Type III (immune complex) reaction

d. Type IV (cell-mediated) reaction

Fill-in-the-Blanks

1. IgE antibodies are _____, meaning they have the ability to bind to target cells such as mast cells.

2. The type of anaphylaxis that develops very rapidly after an antigen is presented to a sensitized host and that may result in shock is _____ anaphylaxis.

3. In the ABO system, an absence of antigens makes a person blood type _____.

4. A graft between identical twins is a(n) _____.

5. The acronym MHC stands for _____.

6. The acronym HLA stands for _____.

7. A sudden drop in blood pressure such as caused by systemic anaphylaxis is called _____.

8. One result of immunosuppression could be development of graft _____ disease.

9. The name of the disease in which an immunological reaction against the thyroid gland receptor sites causes an excessive production of thyroid hormones is _____ disease.

10. The treatment for systemic anaphylaxis is to administer _____ promptly.

11. T cells are involved in _____ hypersensitivity; the reaction takes one or more days to develop.

12. Prevention of hemolytic disease of the newborn in response to Rh incompatibility can be done by administration of _____.

13. An autoimmune condition caused by coating with antibodies the receptor sites at which nerve impulses reach muscles is called _____.

14. A characteristic of true food allergy is appearance of _____ on the skin.

15. Addison's disease is caused by T cells attacking the _____ glands.

163

16. The cornea does not reject transplants; it is an example of a _____.

17. The drug _____ has revolutionized transplantation medicine by selectively suppressing cell-mediated immunity.

18. Pig heart valves are not antigenic and are an example of _____.

19. About 85 percent of the population is Rh _____.

20. Cancer cells escaping immune reactions by shedding specific antigens is called antigen _____.

Key

Matching

I. 1.d 2.b 3.f 4.c 5.a 6.e
II. 1.d 2.d 3.e 4.a 5.g
III. 1.d 2.a 3.c 4.e 5.f
IV. 1.a 2.d 3.c 4.b 5.e
V. 1.d 2.a 3.c 4.d 5.b 6.a

Fill-in-the-Blanks

1. cytotrophic   2. systemic   3. O   4. isograft   5. major histocompatibility complex
6. human leukocyte antigens   7. shock   8. versus host   9. Graves'
10. epinephrine (adrenalin)   11. cell-mediated   12. anti-Rh+ antibodies
13. myasthenia gravis   14. hives   15. adrenal   16. privileged site   17. cyclosporine
18. privileged tissue   19. positive   20. modulation

19 *Antimicrobial Drugs*

Chemotherapy is the treatment of disease with chemicals (drugs) taken into the body. Drugs used for chemotherapy are **chemotherapeutic agents**. The class of chemotherapeutic agents used to treat infectious diseases is **antimicrobial drugs**; unlike disinfectants they must act within the host. **Synthetic drugs** are synthesized in the laboratory; others, **antibiotics**, are produced by microorganisms.

Historical Development

During the early part of this century, Dr. Paul Ehrlich of Germany speculated about a "magic bullet," which would destroy pathogens but not harm the host. Eventually, he found an arsenic derivative, salvarsan, that was useful against syphilis. Prior to this discovery, the ony chemotherapeutic agent available was quinine, for the treatment of malaria. Sulfa drugs were discovered during the 1930s. Penicillin, an antibiotic, was first discovered in 1929 but was not available in a useful form until after 1940.

Criteria for Antimicrobial Drugs

The following criteria can be used to evaluate chemotherapeutic agents:

1. Should demonstrate selective toxicity for microbe but not the host.

2. Should not produce hypersensitivity.

3. Should penetrate body tissues rapidly and be retained for an adequate time.

4. Microbes should not develop resistance to it too readily.

Spectrum of Activity

It is comparatively easy to find antimicrobials against procaryotes (bacteria) because procaryotes differ substantially from the eucaryotic cells of humans. Fungi, protozoans, and helminths are eucaryotic, which makes selective toxicity for the pathogen (without affecting the host) more difficult. It is easily difficult to find antimicrobials against viruses, which exist inside a host cell and interact with the host cell to synthesize new viruses.

   If an antimicrobial drug affects relatively few bacteria it has a narrow spectrum of activity, as opposed to drugs with a broad spectrum of activity. Antibiotics may eliminate normal flora and allow opportunistic pathogens to flourish (**superinfection**).

Action of Antimicrobial Drugs

Inhibition of Cell Wall Synthesis

The cell walls of most bacteria contain peptidoglycans; human cell walls do not. Therefore, interference with the synthesis of these cell walls usually does not harm the host. Antibiotics using this mode of action include penicillins, cephalosporins, cycloserine, bacitracin, and vancomycin. Because the peptidoglycan layer of gram-positive bacteria is more accessible than that of gram-negative, these bacteria are the most susceptible to such agents.

Inhibition of Protein Synthesis

Ribosome structure differs greatly between procaryotic and eucaryotic cells. Many antibodies such as chloramphenicol, gentamicin, erythromycin, tetracyclines, and streptomycin interfere with protein synthesis by reacting with the ribosomes of bacteria.

Injury to the Plasma Cell Membrane

Antibiotics, especially the polypeptides such as polymyxin B, can adversely affect the membrane permeability of microbial cells. Loss of important metabolites occurs from these changes in permeability. Similarly, the effectiveness of nystatin, ketoconazole, and amphotericin B against fungi is based on their combining with sterols to disrupt fungal plasma membranes. The activity of these drugs is selective because animal cell sterols are mostly cholesterol while fungal cells are mostly ergosterol.

Inhibition of Nucleic Acid Synthesis

Similarities between microbial and host cell DNA and RNA are so close that drugs that act by interfering with the nucleic acid synthesis of microbial cells have only limited clinical application. Drugs acting on this principle are rifamycin, nalidixic acid, idoxuridine, and trimethoprim.

Inhibition of Enzymatic Activity

In Chapter 5 we saw how the resemblance of sulfa drugs to para-aminobenzoic (PABA) was responsible for the antimicrobial activity of the former, based upon competitive inhibition. PABA is an essential metabolite for the production of folic acid, which humans do not synthesize. Trimethoprim and sulfones also are antimetabolites.

A summary of the actions of antimicrobial drugs is given in Figure 19-1.

Survey of Commonly Used Antimicrobial Drugs

Synthetic Antimicrobial Drugs

Isoniazid (INH). **Isoniazid**, used in treatment of tuberculosis, is a structural analog of vitamin B_6 (pryidoxine). It is believed to inhibit synthesis of mycolic acids, which are part of the cell wall of mycobacteria.

Ethambutol. **Ethambutol** is effective only against mycobacteria, and is used in chemotherapeutic treatment of tuberculosis. It inhibits the synthesis of certain cellular metabolites.

Sulfonamides. The **sulfonamides** (sulfa drugs) act by competitive inhibiton of folic acid, a precursor to nucleic acids. **Silver-sulfadiazine** is used on burn patients. The most widely used sulfa-containing preparation is the combination of the non-sulfa drug **trimethoprim** and **sulfamethoxazole**. These are structural analogs that inhibit synthesis of DNA at different stages.

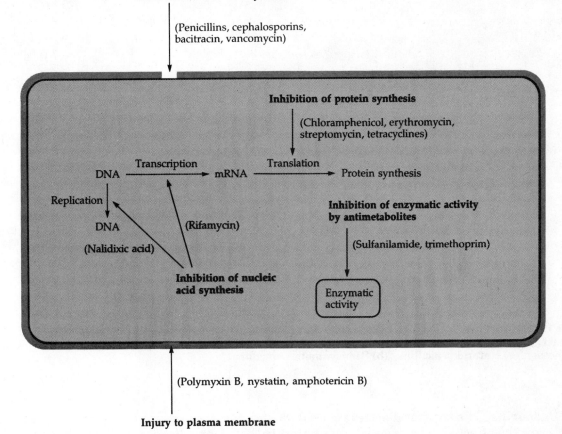

Inhibition of cell wall synthesis

(Penicillins, cephalosporins,
bacitracin, vancomycin)

Inhibition of protein synthesis

(Chloramphenicol, erythromycin,
streptomycin, tetracyclines)

DNA →Transcription→ mRNA →Translation→ Protein synthesis

Replication

DNA

(Nalidixic acid)

(Rifamycin)

**Inhibition of enzymatic activity
by antimetabolites**

(Sulfanilamide, trimethoprim)

**Inhibition of nucleic
acid synthesis**

Enzymatic
activity

(Polymyxin B, nystatin, amphotericin B)

Injury to plasma membrane

Figure 19-1 Actions of antimicrobial drugs in a highly diagrammatic, composite microbial cell. All drugs shown act selectively on procaryotic (bacterial) cells, except for nystatin and amphotericin B, which act on eucaryotic (fungal) cells.

Other synthetic antimicrobial drugs. Nitrofurantoin, of the **nitrofuran family** of drugs, is used against urinary tract infections; **nitrofurazone** and **nifuroxime** are used topically for skin, eye, and vaginal infections. **Nalidixic acid** inhibits DNA synthesis and, because it concentrates in the urine, is used against gram-negative urinary tract infections. The **quinolones** are an emerging group related to nalidixic acid, and have a similar mode of action. The drug **5-fluorocytosine** (Flucytosine) is an antimetabolite of cytosine synthesis and is used mainly as an antifungal drug. Fungal cells convert it to toxic **5-fluorouracil**, and mammalian cells do not.

Antibiotics

Penicillin. The term **penicillin** refers to a group of related antibiotics (see Figure 19-2). Natural penicillins, such as **Penicillin G** or **V**, are products of *Penicillium* mold growth. **Procaine penicillin** and **benzathine penicillin** combine Penicillin G with other drugs to prolong the antibiotic's retention in the body. *Penicillinases* (*beta-lactamases*) are enzymes that cleave the *beta-lactam* ring of penicillins, causing resistance to *semi-synthetic* penicillins such as **ampicillin, amoxicillin**, and **carbenicillin** have a broader spectrum of activity. Penicillinase resistance is a feature of several penicillins such as **methicillin** and **aztreonam**. Potassium clavulanate (clavulanic acid), a non-competitive inhibitor of penicillinase has been combined with broad spectrum penicillin such as **amoxicillin (Augmentin®)** and **ticarcillin (Timentin®)**. Ureidopenicillins such as **mezlocillin** and **azlocillin** have activity against pseudomonads.

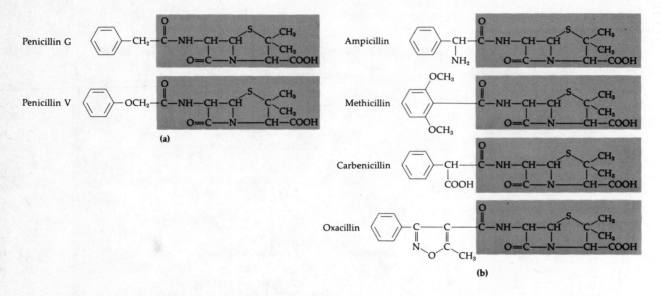

Figure 19-2 Structure of penicillins. The portion that all penicillins have in common, containing the beta-lactam ring, is shaded in color. The unshaded portions represent the side chains that distinguish one penicillin from another. **(a)** Natural penicillins. **(b)** Semisynthetic penicillins.

Cephalosporins. The structural nucleus of **cephalosporins**, the *oxa-beta-lactam ring*, resembles that of penicillin. Cephalosporins such as **cephalothin, cefamandole, cefotaxime**, and **moxalactam** are often used as substitutes for penicillin.

Carbapenems. **Carbapenems** are a new class of very broad spectrum antibiotics. **Primaxin** is a combination of a beta-lactam antibiotic and cilastin sodium which prevents degradation.

Aminoglycosides. **Aminoglycosides** are a group of antibiotics with amino sugars and an aminocyclitol ring. Examples are **streptomycin** (used for tuberculosis treatment; may cause deafness and kidney damage); **neomycin** (used in topical ointment with bacitracin and polymyxin B); and **gentamicin** (effective against most gram-negatives, especially *Pseudomonas*). **Amikacin** is useful in cases of aminoglycoside resistance.

Tetracyclines. **Tetracyclines** are broad-spectrum antibiotics that are also effective against chlamydias and rickettsias. They inhibit protein synthesis. Commonly encountered are **tetracycline, oxytetracycline** (Terramycin), and **chlortetracycline** (Aureomycin). Some newer semisynthetic versions such as **doxycycline** and **minocycline** are retained in the body longer. They produce some toxic side effects such as tooth discoloration and liver damage.

Chloramphenicol.. **Chloramphenicol** is a broad-spectrum antibiotic that affects protein synthesis. Structurally simple, it is often synthesized chemically. It may cause a blood disorder, aplastic anemia. It is the drug choice for typhoid fever and certain types of meningitis where the risk is considered justified.

Macrolides. **Macrolides** are named for their *macrocyclic lactone ring* and are especially effective against gram-positive bacteria. **Erythromycin** inhibits protein synthesis and is used in treating infections resistant to penicillins, as well as mycoplasmal pneumonia and legionellosis.

168

Polypeptides. **Bacitracin** inhibits cell wall synthesis, especially in gram-positives. It is toxic to kidneys and is usually applied topically. **Polymyxin B** injures plasma membranes. It is useful against gram-negative bacteria, especially *Pseudomonas* infections. Because of its toxicity to kidneys and the brain, most applications are topical.

Vancomycin. **Vancomycin** is a toxic drug, apparently unrelated to any other antibiotic, that inhibits peptidoglycan synthesis. It is used against penicillinase-producing staphylococci that cause life-threatening infections.

Rifamycins. **Rifamycins** inhibit RNA synthesis, and are effective against gram-negatives, chlamydias, and poxviruses. Rifampin, the best known, is used in tuberculosis therapy, usually along with INH or ethambutol to minimize drug resistance.

Antifungal Drugs

Polyenes. **Nystatin** and **amphotericin B** are polyene antibiotics whose activity is based on damage to fungal plasma membranes by combining with the membrane sterols. Nystatin is used for local fungal infections to the vagina and skin, such as *Candida* infections. Amphotericin B is used for histoplasmosis, coccidioidomycosis, and blastomycosis (systemic fungal infections).

Imidazoles. **Imidazole** antifungals such as **miconazole** and **clotrimazole** are used topically against cutaneous fungal infections. **Ketoconazole**, taken orally, is a promising substitute for amphotericin B for many systemic fungal infections.

Griseofulvin is a fungistatic drug that seems to interfere with mitosis. Although taken orally, this drug binds selectively to keratin in skin, hair, and nails, and prevents fungal growth at these sites.

Tolnaftate is a topical agent used as an alternative to miconazole for athlete's foot infections.

Antiviral Drugs

Amantadine may interfere with uncoating viruses within the host cell or with transcription of viral RNA. Amantadine may prevent influenza. Examples of purine and pyrimidine structural analogs are **idoxuridine** and **trifluridine**. Both may be used topically for treating herpes keratitis (an eye infection); vidarabine can be used against herpes virus encephalitis. **Ribavirin**, an analog of guanine, has been used in treatment of infant viral pneumonia. **Acylovir** is a nucleoside analog used to treat herpes infections. **Zidovudine, (AZT)** is an analog of thymidine used in treatment of AIDS. **Interferon** (see Chapter 15) has become more available from recombinant DNA manufacturing techniques and may prove useful as an antiviral agent.

Antiprotozoan and Antihelminthic Drugs

Quinine still has limited use against malaria, but has been replaced with synthetic derivatives such as **chloroquine**. Chloroquine fits between base pairs in DNA (*intercalation*) and prevents DNA synthesis. **Quinacrine**, used against giardiasis, functions similarly. **Diiodohydroxyquin** is an amoebicide. **Metronidazole** is used for treatment of many protozoan diseases and also is effective against certain anaerobic bacteria. It probably causes disruption of DNA under anaerobic conditions. **Pentamidine isethionate** is used for treatment of *Pneumocystis* pneumonia. **Nifurtimox** is used to treat Chagas' disease, caused by a trypanosome. **Niclosamide** inhibits oxidative phosphorylation helminths. **Stibophen** is an *antimonial* anti-helminithic drug. **Praziquantel** also is effective against tapeworms and several fluke-caused diseases. **Mebendazole** is used to treat ascariasis. **Piperazine** paralyzes pinworms, which are then passed out of the body.

Tests for Microbial Susceptibility to Chemotherapeutic Agents

The Disk Diffusion Method

The **disk diffusion method** (**Kirby-Bauer** test) uses a dish of agar medium seeded uniformly with a test organism. Filter paper disks impregnated with known concentrations of chemotherapeutic agents are placed on the agar surface. If the chemotherapeutic agent is effective, a zone of inhibition (no growth) is observed around the disk. The diameter of the zone can be used to calculate the susceptibility of the organisms to the agent.

Tube Dilution and Agar Dilution Tests

A series of dilutions of an antibiotic can be placed in tubes (shallow wells in a plastic plate usually are used in practice) and inoculated with test bacteria. After incubation they are examined for turbidity. The **minimum inhibitory concentration** (**MIC**) of the antimicrobial is defined as the lowest concentration that prevents growth. Subculturing from the tubes that show no growth will determine if the bacteria have been killed or only inhibited. The lethal concentration that actually kills the bacteria is called the **minimum bactericidal concentration** (**MBC**). Many of these tests are highly automated and use light-scattering to determine bacterial growth.

Effectiveness of Chemotherapeutic Agents

Drug Resistance

Resistance to drugs may be based on production of an enzyme such as penicillinase. Another serious threat to chemotherapy is drug resistance of various kinds that is carried on **plasmids**, called **resistance (R) factors**. Plasmids may transfer genes for resistance to several antibiotics at a time and to closely related bacterial species. To help avoid development of resistance, drugs should be used with discrimination. Simultaneous administration of more than one antibiotic helps avoid development of resistance. When two drugs are used simultaneously, there also may be a synergistic effect in which the results are superior to those of either drug used by itself. Other combinations of drugs may be **antagonistic**, meaning that together they do not work as well as they do independently.

There is a pressing need for better antiviral drugs. Because of the similarity between requirements for cancer therapy and requirements for viral disease treatment, perhaps research for one will benefit the other.

Test

In the matching exercises there is only one answer to a question; however, the lettered options (a, b, c, etc.) may be used more than once or, perhaps, not at all.

I. Match

1. Plasmids that carry antibiotic resistance.

2. Chemotherapeutic agents produced by micro-organisms.

3. Sulfa drugs are an example of this type of antibiotics.

    a. Antibiotics
    b. Synthetic drugs
    c. Chemotherapy
    d. Kirby-Bauer test
    e. R factors

II. Match

1. Combines with sterols in plasma membrane.

2. Inhibition of protein synthesis.

3. Inhibition of RNA synthesis.

4. Inhibition of synthesis of cell-wall peptidoglycans.

5. Inhibition of DNA synthesis.

6. Inhibition of cell-wall mycolic acids.

    a. Cephalosporins
    b. Chloramphenicol
    c. Nystatin
    d. Isoniazid
    e. Sulfonamides
    f. Rifamycins

III. Match

1. Contains oxy-beta-lactam ring in structure.

2. Synthetic drug used in tuberculosis chemotherapy.

3. Antifungal.

4. Causes plasma membrane leakage; useful against *Pseudomonas*.

5. Antiviral drug; a pyrimidine analog.

    a. Cephalosporin
    b. Polymyxin B
    c. Ethambutol
    d. Idoxuridine
    e. Nystatin

IV. Match

1. Used in treating diseases caused by protozoans.

2. Antifungal drug taken orally that concentrates in keratin.

3. Useful against tapeworms.

4. A drug that is useful against symptoms of genital herpes.

5. An antifungal drug of the polyene type.

6. A macrolide antibiotic.

a. Erythromycin

b. Griseofulvin

c. Amphotericin B

d. Niclosamide

e. Metronidazole

f. Acylovir

V. Match

1. Inhibits oxidative phosphorylation in helminths.

2. Interferes with DNA replication by intercalation.

3. Used in treatment of Pneumocystis pneumonia.

4. Paralyzes pinworms.

5. Treatment of Chagas' disease.

a. Chloroquine

b. Niclosamide

c. Pentamidine isethionate

d. Piperazine.

e. Nifurtimox

VI. Match

1. Used in treatment of herpes encephalitis.

2. Synthetic antimicrobial that blocks synthesis of nucleic acids.

3. Derivative of Penicillin G that is designed to be retained for a longer time in the body.

4. A penicillin designed to be resistant to penicillinase.

5. A nitrofuran used topically against skin, eye, and vaginal infections.

6. An amoebicide.

7. Very broad spectrum; carbapenem group.

a. Adenine arabinoside

b. Diiodohydroxyquin

c. Methicillin

d. Ampicillin

e. Procaine penicillin

f. Trimethoprim

g. Nifuroxime

h. Primaxin

Fill-in-the-Blanks

1. Cell walls of bacteria contain _____, and human cells do not.

2. The treatment of disease with chemicals taken into the body is called _____.

3. Nystatin and amphotericin B disrupt fungal plasma membranes by combining with _____.

4. Nalidixic acid inhibits _____ synthesis.

5. 5-fluorocytosine is a structural analog of _____.

6. The first "magic bullet" was an arsenical derivative called _____, which was useful against syphilis.

7. Many bacteria develop resistance to penicillin by producing the enzyme _____.

8. An analog of vitamin B_6 used in tuberculosis therapy is _____.

9. The drug of choice in treating typhoid fever may sometimes cause a blood condition called _____.

10. The drug 5-fluorocytosine is converted by fungi into the toxic form 5-fluoro _____.

11. An aminoglycoside antibiotic used in tuberculosis treatment, but that may cause deafness and kidney damage is _____.

12. The term *penicillin* is applied to a group of antibiotics that all have a _____ in their structure.

13. The aminocyclitol ring and amino sugars are found in the _____ group of antibiotics.

14. A very toxic antibiotic, _____, is not related to any other antibiotic and is used only against life-threatening staphylococcal infections resistant to penicillin.

15. The lowest concentration of chemotherapeutic agent that will prevent growth is the _____.

16. The lowest concentration of a chemotherapeutic agent that will kill the pathogen (as contrasted to inhibition) is called the _____.

17. Miconazole and ketoconazole are examples of _____ type antifungals.

18. Many antibiotics inhibit protein synthesis by reacting with the _____ of the bacterium, which differ greatly between procaryotic and eucaryotic cells.

19. When antibiotics eliminate much of the natural flora there may be an overgrowth of resistant pathogens; this is called _____.

20. _____ is a synthetic quinine derivative used in the treatment of malaria.

21. Zidovudine, or _____, is an analog of thymidine and is used in treatment of AIDS.

Key

Matching

1. 1.e 2.a 3.b
II. 1.c 2.b 3.f 4.a 5.e 6.d
III. 1.a 2.c 3.e 4.b 5.d
IV. 1.e 2.b 3.d 4.f 5.c 6.a
V. 1.b 2.a 3.c 4.d 5.e
VI. 1.a 2.f 3.e 4.c 5.g 6.b 7.h

Fill-in-the-Blanks

1. peptidoglycan   2. chemotherapy   3. sterols   4. DNA   5. cytosine   6. salvarsan
7. penicillinase   8. isoniazid (INH)   9. aplastic anemia   10. -uracil
11. streptomycin   12. beta-lactam ring   13. aminoglycoside   14. vancomycin
15. minimum inhibitory concentration (MIC)
16. minimum bactericidal concentration (MBC)   17. imidazole   18. ribosomes
19. superinfection   20. chloroquine   21. AZT

20 *Microbial Diseases of the Skin and Eyes*

The first lines of defense of the body are the skin and the mucous membranes.

Structure and Function of the Skin

The **epidermis** is the outer, thinner portion of the skin; it is composed of several layers of epithelial cells. The outer epidermal layer consists mainly of dead cells containing the protein **keratin**. The inner layer of the skin is the **dermis**, composed of connective tissue with numerous blood and lymph vessels, nerves, hair follicles, and sweat and oil glands. Perspiration provides the moisture necessary for microbial growth; however, salts interfere with the growth of many organisms, and lysozyme in perspiration digests the cell walls of many bacteria.

Sebum from oil glands is mainly a mixture of unsaturated fatty acids, proteins, and salts. The fatty acids inhibit the growth of certain pathogens. Sebum helps prevent drying of the skin and hair but is nutritive to certain microorganisms, as will be discussed later. In the lining of body cavities such as the mouth, nasal passages, urinary and genital tracts, and gastrointestinal tract, there is a layer of specialized epithelial cells, the **mucous membrane (mucosa)**. Some cells produce **mucus**, which traps particles, including microbes. Other cells are ciliated, functioning to sweep particles and mucus from the body. The eyes are washed by tears, and lysozyme in tears destroys many bacteria.

Normal Flora of the Skin

Bacteria tend to clump on the skin in small groups. Many are gram-positive spherical bacteria such as *Staphylococcus* and *Micrococcus*. These often produce antimicrobial substances. Gram-positive pleomorphic rods, **diphtheroids**, such as *Propionibacterium acnes* metabolize the oil secretions in hair follicles. Acid produced by them keeps the skin pH between 3 and 5, a condition that tends to be bacteriostatic. Other diphtheroids such as *Corynebacterium xerosis* are aerobic. The yeast *Pityrosporum* grows on oily skin secretions. High moisture regions such as the axilla (armpit) are moist enough to support the growth of microorganisms, and populations there are high compared to dry regions such as the scalp.

Bacterial Diseases of the Skin

Staphylococcal Skin Infections

Staphylococci are spherical gram-positive bacteria that form irregular grapelike clusters of cells. *Staphylococcus aureus* is the most pathogenic of the staphylococci. It forms golden yellow colonies on agar, and almost all pathogenic strains product **coagulase**, an enzyme that coagulates the fibrin in blood. The coagulase test is used to distinguish *S. aureus* from other species of *Staphylococcus*. *S. aureus* produces a number of toxins, such as enterotoxins, that affect the gastrointestinal tract (which will be discussed in Chapter 24), **leukocidins**, which destroy phagocytic leukocytes, and **exfoliative toxin**, which is responsible for scalded skin syndrome. About 30 percent of the human population carries potentially pathogenic *S. aureus*. It is found primarily in nasal passages but is well adapted, because of resistance to environmental stresses there, to life on the skin. *S. aureus* is a common hospital problem. It is universally present; constant exposure to antibiotics causes rapid development of resistance. The species, therefore, causes many infections of surgical wounds and other artificial breaches of the skin barrier.

Pimples are infections of hair follicles, which are natural openings in the skin barrier; an eyelash follicle infection is a sty. **Furuncles (boils)** are a type of abscess usually arising from a hair follicle infection. In furuncles, a region of pus is surrounded by inflamed tissue—more extensive invasion of tissue is termed a **carbuncle**. Established abscesses do not respond well to antibiotic therapy because the bloodstream does not reach them. When abscess-causing organisms gain access to the bloodstream and increase rapidly in number, **septicemia** results.

Impetigo of the newborn is troublesome in hospital nurseries and prevention of this staphylococcal infection is based largely on the use of hexachlorophene-containing skin lotions. When the outer skin layers peel away in reaction to staphyloccal toxins, it is known as the **scalded skin syndrome.** This syndrome is one of the characteristics of **toxic shock syndrome**, caused by growth of staphylococcal bacteria, especially in the vagina. The **Thucydides** syndrome is a staphylococcal pneumonia associated with influenza.

Streptococcal Skin Infections

Streptococci are gram-positive spherical bacteria that tend to grow in chains. They are aerotolerant anaerobes that do not use oxygen and are catalase-negative. They produce several toxins, as well as enzymes such as **hemolysins**. Beta-hemolytic group A streptococci such as *Streptococcus pyogenes* are the most important from the standpoint of human disease. The *M protein* distributed on the surface of their fimbriae (which is also the basis for subdividing them into over 55 immunologic types) contributes to pathogenicity by its antiphagocytic properties and as an aid in adherence. Also produced by these cells are *erythrogenic toxins* (scarlet fever rash), *deoxyribonucleases* (enzymes degrading DNA), NAD*ase* (an enzyme that breaks down NAD), *streptokinases* (enzymes dissolving blood clots), *hyaluronidase* (an enzyme that dissolves hyaluronic acid, the fluid substance of connective tissue), and *leukocidins* (enzymes that kill white blood cells). Beta-hemolytic streptococci are differentiated into immunologic groups A through O based on carbohydrate antigens.

Impetigo is a superficial skin infection affecting grade school age children; this infection is characterized by pustules (small, round elevations containing pus) that become crusted and rupture. The bacteria enter through minor lesions. Staphylococci often are present but are probably secondary invaders. Penicillin and erythromycin are effective in treatment.

S. pyogenes is the most common cause of **erysipelas**, a disease in which the dermis is affected and the skin erupts into reddish patches. Usually it is preceded by a streptococcal infection, such as a sore throat, elsewhere in the body. Penicillin and erythromycin are used in treatment.

Infections by Pseudomonads

Pseudomonads, particulary *Pseudomonas aeruginosa*, may cause opportunistic skin infections. These bacteria are common in soil, water, and plants, and grow on minimal organic material. Resistance to antibiotics often makes them a hospital problem. *P. aeruginosa* produces several exotoxins and endotoxins. It may cause respiratory infections in compromised hosts (especially cystic fibrosis) or in burn patients, where infection may produce a characteristic blue-green pus from *pyocyanin pigments*. **Otitis externa**, or swimmer's ear, is a pseudomonad infection of the outer ear. Skin infections, ***Pseudomonas dermatitis***, are contracted in such places as whirlpool baths and the like—the organisms are relatively resistant to chlorines. Gentamicin and carbenicillin, often used in combination, are effective; silver sulfadiazine is used to treat *P. aeruginosa* infections of burn patients.

Acne

Acne is probably the most common skin disease of humans. It begins when sebum accumulates, forming whiteheads, and later blackheads. Bacteria such as staphylococci, and the diphtheroids such as *P. acne* become involved if the accumulation ruptures the hair follicle lining. *P. acne* metabolizes sebum, forming free fatty acids that cause the inflammation leading to scarring. Diet, including consumption of chocolate, has no significant effect on incidence of the disease. Benzoyl peroxide is useful, and the tetracyclines also have been used in treatment. Because isotretinoin (Accutane®) inhibits sebum formation, it is used to treat severe cases of **cystic acne**.

Viral Diseases of the Skin

Warts

Some papovaviruses cause the uncontrolled nonmalignant growth of skin cells called warts. Warts can be spread by contact; the incubation period is several weeks. Treatment of warts includes burning them with acids, or the use of liquid nitrogen cryotherapy or electrodesiccation. Genital warts have been successfully treated with injected interferon.

Smallpox (Variola)

Smallpox is caused by a poxvirus known as the smallpox (variola) virus. Variola major smallpox has a mortality rate of 20 percent or more, and variola minor has a mortality rate of less than 1 percent. Recovery from one gives immunity to either. Transmission is by the respiratory route, and the viruses eventually move through the bloodstream (viremia) to infect the skin. Vaccination has resulted in the elimination of smallpox in the world.

Chickenpox (Varicella) and Shingles (Herpes Zoster)

Chickenpox (varicella) is a relatively mild disease, the second most reportable disease (after gonorrhea) in the United States. Infection is acquired by the respiratory route, and after about two weeks incubation, it localizes in the skin. The skin is vesicular for 3 or 4 days. The herpes virus that causes chickenpox may remain latent in nerve cells and later in life be reactivated to cause **herpes zoster** or shingles. In shingles, vesicles similar to chickenpox occur in areas such as the waist, upper chest, and face, affecting nerve branches of the cutaneous sensory nerves.

Reye's syndrome is an occasional, severe complication of chickenpox, influenza, and sometimes other viral diseases. Brain damage or death may result from brain swelling, which prevents blood circulation. Aspirin may increase chances for the Reye's syndrome complication.

Herpes Simplex

The herpes simplex virus is able to remain latent for long periods of time following an initial infection, which usually occurs in infancy and affects the oral mucous membrane. Infections recur in the form of **cold sores** or **fever blisters** when triggered by stress at a later time. These infections are due to **herpes simplex type 1 virus**. Another infection caused by a similar virus, **herpes simplex type 2 virus**, is transmitted primarily by sexual contact and will be discussed in Chapter 25.

Measles (Rubeola)

Measles is spread by the respiratory route and is caused by a paramyxovirus. Because there is no real reservoir other than humans, measles could potentially be eradicated. Measles is extremely dangerous; complications such as encephalitis, which may result in brain damage, occur once in each 2000 cases, and once in each 3000 cases there is a fatality. Complications such as middle ear infections and pneumonia are also common. Ten to twelve days after infection, a papular rash appears.

Koplik spots on the oral mucosa are diagnostically useful. Measles viruses may remain latent and later lead to autoimmune reactions such as nerve destruction in **subacute sclerosing panencephalitis (SSPE)**. The vaccine is usually of the trivalent type of measles, mumps, and rubella (MMR).

Rubella

Rubella is caused by a togavirus (rubella virus), and is much milder than rubeola, often being subclinical. There is a light fever and a rash of small red spots. Transmission is by the respiratory route. Infections during the first trimester of pregnancy can lead to the **congenital rubella syndrome**. This causes serious defects such as deafness, eye cataracts, hearing defects, and mental retardation. Immunoglobulin for passive immunization against rubella is recommended for exposed persons living in the same household as a pregnant woman. Diagnostic tests are hemagglutination inhibiton or ELISA (Chapter 17).

Fungal Diseases of the Skin

A fungal infection of the body is referred to as a mycosis. The ability of fungi to resist high osmotic pressure and low moisture makes the skin susceptible to fungal infections.

Superficial Mycoses

Superficial mycoses called **piedras** are limited to the external shafts of hair.

Cutaneous Mycoses

Fungi that colonize hair, nails, and the outer layer of the epidermis are **dermatophytes**. **Tinea nigra**, caused by *Clasdosporium werneckii*, causes brown or black blotches on the palms of the hands. **Ringworm**, or **tinea**, is caused by members of the genera *Microsporum*, *Trichophyton*, and *Epidermophyton*. **Tinea capitis**, or ringworm of the scalp, is spread by contact among children and by contact with various fomites, dogs, and cats. Ringworm of the groin (jock itch) is **tinea cruris** and ringworm of the feet (athlete's foot) is **tinea pedis**. Hair and nails, high in keratin, are areas for which fungal dieseases have an affinity. Topical agents for tinea infections are miconazole, clotrimazole, tolnaftate, and zinc undecylenate. An oral antibiotic, griseofulvin, localizes in keratinized tissue and is an effective treatment.

Subcutaneous Mycoses

Subcutaneous mycoses are caused by fungi inoculated into a wound. The nodule formed by the growing fungus may spread along the lymphatic vessels. **Chromomycosis** is characterized by lesions that are pigmented dark brown; it is caused by black molds such as *Philophora verruscosa* and *Fonsecaea pedrosoi*. Flucytosine is an effective treatment. **Maduromycosis** (also called mycetoma) is caused by *Allescheria boydii*. Surgical drainage is important, and polyene antibiotics may be useful in treatment.

Candidiasis

Candida albicans is a yeastlike fungus that causes infections of the mucous membranes of the genitoourinary tract (**vaginitis**) andoral mucosa (**thrush**). These infections, collectively called **candidiasis**, often appear when bacterial populations are suppressed by antibiotics that do not affect fungi. For tropical use, nystatin, clotrimozole, or miconazole are recommended.

Diseases of the Eye

Bacterial Diseases of the Eye

Conjunctivitis is an inflammation of the **conjunctiva**, the mucous membrane that lines the eyelid and covers the outer surface of the eyeball. **Contagious conjunctivitis, pinkeye**, is caused by *Hemophilus aegyptius*. **Neonatal gonorrheal ophthalmia** is a dangerous eye inflammation transmitted to the newborn during passage through the birth canal. Newborn infants have their eyes washed with 1 percent silver nitrate or antibiotics to prevent this infection.

Trachoma is the greatest single cause of blindness in the world. It is an infection of the epithelial cells of the eye caused by *Chlamydia trachomatis*. Scar tissue forms on the cornea. It is transmitted by contact with fomites. In the early stages, sulfonamide and tetracycline are useful in treatment. **Inclusion conjunctivitis** is caused by the same organisms but involves only the conjunctiva, not the cornea. This disease may be spread (to newborns) by an infection in the birth canal or by unchlorinated waters. Tetracycline is used in treatment.

Viral Infections

Herpetic keratitis is caused by herpes simplex type 1 virus, with an epidemiology similar to cold sores. Corneal ulcers occur and may lead to blindness. Trifluridine is the drug of choice against this viral infection.

Protozoan Diseases of the Eye

A freshwater amoeba, *Acanthamoeba*, causes a keratitis in the eyes of contact lens wearers. The damage is severe and may require a corneal transplant to correct.

Test

In the matching exercises there is only one answer to a question; however, the lettered options (a, b, c, etc.) may be used more than once or, perhaps, not at all.

I. Match

1. Inner layer of the skin, composed of connective tissue.

2. Lining of the inner eyelid and the surface of the eyeball.

3. Some of these specialized epithelial cells are ciliated.

4. Consists largely of dead cells containing the protein keratin.

a. Mucous membrane

b. Epidermis

c. Dermis

d. Conjunctiva

II. Match

1. *Streptococcus pyogenes.*

2. *Staphylococcus aureus.*

3. Tinea.

4. *Propionibacterium acne.*

5. Parvovirus.

a. Scalded skin syndrome

b. Acne

c. Erysipelas

d. Warts

e. Ringworm

III. Match

1. Variola.

2. Varicella.

3. Herpes zoster.

4. Rubeola.

5. Rubella.

6. Shingles.

a. Impetigo

b. Smallpox

c. Measles

d. Chickenpox

e. German measles

Key

Matching

1. 1.c 2.d 3.a 4.b
II. 1.c 2.a 3.e 4.b 5.d
III. 1.b 2.d 3.d 4.c 5.e 6.d
IV. 1.a 2.a 3.c 4.d
V.   1.c 2.d 3.a 4.b
VI. 1.c 2.b 3.a 4.e 5.d

Fill-in-the-Blanks

1. lysozyme 2. diphtheroids 3. Reye's 4. carbuncle 5. beta 6. pyocyanin
7. acne 8. measles (rubeola) 9. mycosis 10. mycetoma 11. trachoma 12. 1
13. sty 14. antiphagocytic 15. sebum 16. smallpox 17. trimester
18. capitis 19. coagulase 20. erythrogenic 21. pedis 22. *Acanthamoeba*
23. influenza

21 *Microbial Diseases of the Nervous System*

Structure and Function of the Nervous System

The human nervous system consists of billions of nerve cells (neurons). The **central nervous system (CNS)** consists of the **brain** and **spinal cord**. It is the control center that picks up sensory information from the environment and, after interpreting it, sends out impulses to coordinate body activities. The **peripheral nervous system** consists of all the nerves branching from the brain and spinal cord. These nerves are the communication lines between the CNS and the body, and the external environment.

The skull protects the brain and the backbone protects the spinal cord; both the brain and spinal cord are covered by the **meninges** (Figure 21-1). The meninges consist of the dura mater (outermost layer), the arachnoid (middle layer), and the pia mater (innermost layer). **Cerebrospinal fluid** circulates in spaces within the brain (ventricles) and the space between the pia mater and the arachnoid layers (subarachnoid space). The **blood-brain barrier** consists of capillaries that permit certain substances to pass from the blood to the brain but prevent others. Microorganisms can gain access to the CNS by trauma (accidental, or medical procedures such as spinal taps). Also, microorganisms may enter by movement along peripheral nerves or by the bloodstream and lymphatic system. An infection of the meninges is **meningitis**; an infection of the brain is **encephalitis**. Antibiotics often are unable to cross the blood-brain barrier.

Bacterial Diseases of the Nervous System

Bacterial Meningitis

The three major types of bacterial meningitis are: meningococcal meningitis, caused by *Neisseria meningitidis*; pneumococcal meningitis, caused by *Streptococcus pneumoniae*; and *Hemophilus influenzae* meningitis. At least 50 other bacteria can cause meningitis, as can fungi, viruses, and protozoans.

Neisseria Meningitidis (Meningococcal Meningitis)

Meningococcal meningitis (cerebrospinal fever) is caused by the bacterium *Neisseria meningitidis*, an aerobic, nonmotile, gram-negative coccus. It is a frequent inhabitant of the throat behind the nose, and a throat infection can lead to bacteremia followed by meningitis. Symptoms are thought to be due to an endotoxin produced by the bacteria. Meningitis symptoms begin much as a mild cold, followed by fever, severe headache, and neck and back stiffness. Neurological complications such as convulsions, deafness, blindness, and minor paralysis are not uncommon—in severe cases, death can occur in a few hours. The endotoxin causes extensive blood vessel damage and heightens sensitivity to further exposure to endotoxin. The increased sensitivity is called the Schwartzman phenomenon.

The disease primarily affects the very young, with the highest incidence in the first year. Before development of an effective vaccine used by the military, outbreaks among young recruits were common. The organism is relatively resistant to phagocytosis. The meninges are not penetrated by most antibiotics. Sulfas penetrate well but resistance is a problem. Diagnosis is based on isolation of bacteria

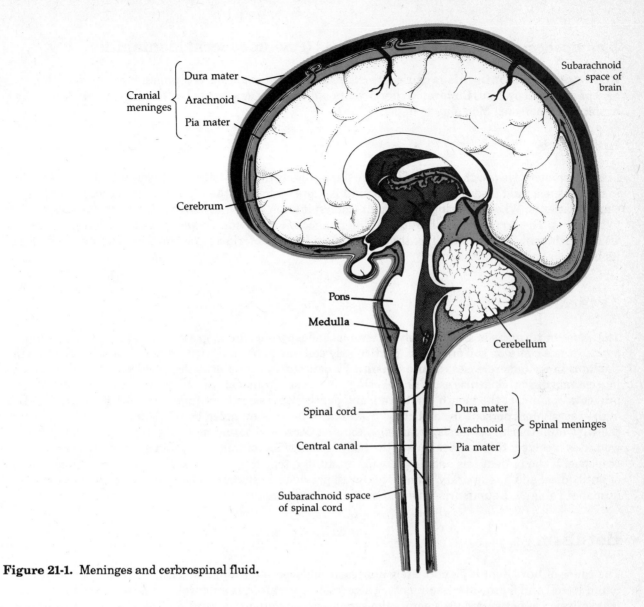

Figure 21-1. Meninges and cerbrospinal fluid.

from blood or cultures of lumbar punctures. For isolation, a candle jar or CO_2 incubator to produce an environment of 5 to 10 percent CO_2 is useful. Countercurrent immuno-electrophoresis of cerebrospinal fluid is useful in differentiating between bacterial causes of meningitis.

Hemophilus Influenzae Meningitis

Hemophilus influenzae type b, the only serological type of medical importance, is an aerobic, gram-negative pleomorphic bacterium. A member of the normal throat flora, it also can cause meningitis. The name derives from the need for X and V factors from lysed red blood cells and the erroneous idea at one time that it was the cause of influenza. It is the most common cause of bacterial meningitis among children under 4 years of age. Following a viral infection of the respiratory tract, *H. influenzae* enters the blood and then invades the meninges. Laboratory diagnosis is based on analysis of spinal fluid and blood for gram-negative organisms or antibodies. A passive agglutination test using latex particles is now available.

Streptococcus pneumoniae Meningitis (Pneumococcal Meningitis)

Pneumococcal meningitis is caused by *Streptococcus pneumoniae*, a common inhabitant of the nasopharyngeal system. Children under 5-years-old are very susceptible to this disease, as are hospitalized persons. Mortality is high.

Listeriosis

Listeria monocytogenes, the cause of **listeriosis**, is widely distributed, mainly in animal feces. Listeriosis is a mild disease in most adults but is more serious in the immunosuppressed, including cancer patients. When infecting a pregnant woman it often causes abortion, or serious damage or death to the fetus. The organism proliferates within phagocytic cells, and grows at refrigeration temperatures. Cold enrichment, in which the fact that it outgrows other bacteria at low temperatures has been used for its isolation and identification

Tetanus

The cause of **tetanus** is an obligately anaerobic, endospore-forming, gram-positive rod, *Clostridium tetani*. It is a common soil organism, particularly soil contaminated with animal fecal wastes. Growth of *C. tetani* in wounds releases **tetanospasmin**, a neurotoxin that blocks the nerve pathway that signals muscle relaxation. Spasms result as opposing muscle sets contract. Contraction of the jaw muscles prevents opening of the mouth (**lockjaw**), and death from respiratory failure results in about one-third of the cases. Most persons in the United States have been immunized by the tetanus toxoid included in the DPT (diphtheria, pertussis, tetanus) vaccine. A booster of toxoid may be given when a dangerous wound is received; this booster causes the body to renew its immunity level in a rapid anamnestic response. If the patient has had no previous immunity, toxoid may not cause a rapid enough appearance of antibodies, and a temporary immunity can be provided by immune globulin pooled from immunized humans. Table 21-1 summarizes wound management for tetanus.

Botulism

The cause of **botulism** is an obligately anaerobic endospore-forming bacterium, *Clostridium botulinum*, found in soil and freshwater sediments. The exotoxin produced is a neurotoxin that blocks the transmission of nerve impulses across the synapses, causing progressive *flaccid paralysis*. Symptoms typically appear in a day or two. The word *botulism* is derived from the Latin, *botulus*, meaning

---

Table 21-1. Wound Management for Prevention of Tetanus

History of Tetanus Immunization	Minor Wounds		Extensive Wounds	
	Administer Toxoid	Administer TIG†	Administer Toxoid	Administer TIG†
Unknown or 0 to 2 doses	Yes	No	Yes	Yes
3 or more doses	No*	No	No**	No

†TIG = tetanus immune globulin
*Yes, if more than 10 years since last dose.
**Yes, if more than 5 years since last dose.

"sausage," a common vehicle of the disease at one time. Most botulism results from attempts at preservation by heat that fail to eliminate the *C. botulinum* endospore, but provide anaerobic conditions for its growth. There are several toxin types. **Type A toxin** is found mostly in the western United States. It is proteolytic and probably the most virulent. **Type B toxin** is less virulent, and both proteolytic and nonproteolytic strains occur. Type B is responsible for most outbreaks in the eastern United States and Europe. **Type E toxin** is produced by a strain found in wet soils and sediments, and often involves seafood. This organism is nonproteolytic, its endospores are less heat-resistant, and the toxin can be produced at refrigerator temperatures.

The botulinal toxin is heat-labile; that is, it is destroyed by boiling. It is not formed in acid foods below pH 4.7. Nitrites are included in some meat products to prevent bacterial growth following germination of endospores. There is some concern, however, that nitrites are responsible for formation of **nitrosamines**, a carcinogen. For treatment, antitoxins of trivalent ABE type are available, but they do not affect attached toxins. Toxin identification can be done by inoculating mice with the suspected toxin; if the mice who survive the inoculation are protected by, for example, type A antitoxins, then the suspected toxin is considered to be type A.

Wound botulism can occur from *C. botulinum* growth in wounds, and **infant botulism** may occur from growth in the infant GI tract, which is rarely a factor in adults.

Leprosy

Leprosy, sometimes called Hansen's disease for the person who first isolated the organism, is caused by an acid-fast rod, *Mycobacterium leprae*. It is probably the only bacterium that grows primarily in the peripheral nervous system. The microorganisms have not been grown on artificial media but can be grown on the footpads of mice and in armadillos. There are two main forms of leprosy: the **tuberculoid (neural) form** is characterized by regions of the skin that have lost sensation and are surrounded by a border of nodules. The lepromin test, similar to the tuberculin test in design, usually is positive for this form. Tuberculoid leprosy occurs in persons with an effective immune system; in the **lepromatous (progressive) form** of leprosy, skin cells are infected, and disfiguring nodules form over the body. Mucous membranes of the nose (one sign is a lion-faced appearance) and the hands are particularly affected—the organism prefers cooler regions of the body.

Leprosy is not very contagious. Death usually is a result of complications from diseases such as tuberculosis. Laboratory diagnosis is by identification of acid-fast rods in lesions. The sulfone drugs such as dapsone are the most effective treatment. Rifampin and a fat-soluble dye, clofazimine, also are used.

Viral Diseases of the Nervous System

Poliomyelitis

Poliomyelitis is informally called polio. Adults are more likely to get the disease when living in regions of good sanitation, where immunity is not acquired as an infant; therefore, in adults, the disease is more likely to reach the paralytic stage. Symptoms usually are a few days of headache, sore throat, and fever, and only in a few cases are paralytic. The poliovirus occurs in three different serotypes. The primary mode of transmission is ingestion of fecal-contaminated water. In paralytic cases, the virus penetrates capillary walls and enters the CNS, where it displays a high affinity for nerve cells. The motor nerve cells in the upper spinal cord, called the anterior horn cells, are particularly affected.

Two vaccines are available; the **Salk vaccine** uses viruses inactivated by formalin. This vaccine requires a series of injections and periodic booster shots; the **Sabin vaccine** contains three live attenuated strains. The immunity from this vaccine resembles that acquired naturally, but on rare occasions the vaccine itself may cause the disease, probably by a mutation to virulence. Diagnosis usually is based on isolation of the virus and observation of cytopathic effects on cell cultures. Viral neutralization serological tests also can be found.

Rabies

The **rabies virus**, a rhabdovirus with a bullet shape, is typically transmitted in the saliva of a biting rabid animal. The virus travels along peripheral nerves to the CNS where it produces encephalitis. Symptoms include spasms of the muscles of the mouth and pharynx when swallowing liquids. This painful reaction causes an aversion to water, which is the basis of the term **hydrophobia** (fear of water) that is applied to rabies.

Dogs with **furious rabies** are restless, snapping at anything. Some animals suffer from **dumb (paralytic) rabies** in which there is only minimal excitability although they may snap irritably if handled.

Laboratory diagnosis usually is made by a fluorescent antibody test. In early years, the Pasteur treatment was standard for prevention of rabies in exposed individuals. The vaccination procedure consisted of 14 to 21 injections of brain tissue of rabbits that died from rabies virus. This tissue had been inactivated by physical or chemical means before injection. Currently, treatment of exposed persons begins with the administration of human rabies immune globulin (RIG) followed by active immunization. The latter may be done with rabies virus grown in duck embryo culture (duck embryo vaccine, or DEV) or with human diploid cell vaccines (HDCV) grown in human diploid cell lines. HDCV is considered preferable, as it is more effective and less likely to produce allergic reactions. HDCV is administered in five injections over a four-week period, and a sixth dose is given two months later. Indications for antirabies treatment are unprovoked bites from dogs, cats, skunks, and similar animals. Bites from rodents and rabbits generally do not require antirabies treatment. It also is possible to contract rabies by inhalation of aerosols of the virus.

Arthropod-Borne Encephalitis

Encephalitis, caused by arboviruses that are transmitted by a mosquito (of the genus *Culex*), is common in the United States. Horses frequently are infected, accounting for terms such as **Eastern equine** and **Western equine, (EEE, WEE)**.**St. Louis encephalitis (SLE)** and **California encephalitis**, which do not infect horses, also occur. In the Far East, **Japanese B encephalitis** is a similar disease. All cause a percentage of neurological damage; however, EEE is the most severe, causing a high incidence of brain damage, deafness, and similar neurological problems. It also has a higher mortality rate. Diagnosis usually is made by complement fixation tests. Control of *Culex* mosquitoes is an essential control measure.

Fungal Disease of the Nervous System

Cryptococcus neoformans Meningitis (Cryptococcosis)

Cryptococcus neoformans is a yeastlike fungus widely distributed in soil, especially in soil enriched by pigeon droppings. The disease **cryptococcosis** is thought to be transmitted by inhalation of dried infected pigeon droppings. Often the disease does not proceed beyond an infection of the lungs, but it may spread by the bloodstream to the brain and meninges. Laboratory diagnosis is by identification of the microorganisms in an India ink wet mount and observation of virulence in mice. The drug of choice for treatment is amphotericin B or flucytosine.

Protozoan Diseases of the Nervous System

African Trypanosomiasis

African trypanosomiasis is a disease caused by protozoans that affects the nervous system. The flagellates *Trypanosoma brucei gambiense* and *Trypanosoma brucei rhodensiense* enter the body through a tsetse fly bite. Domestic and wild animals are reservoirs for the disease. Later they move into the cerebrospinal fluid, where the resulting symptoms are responsible for the informal name of sleeping sickness. Suramin, pentamidine, and a toxic arsenical called melarsoprol are used in chemotherapy.

Naegleria Microencephalitis

An amoebic protozoan found in ponds and streams, *Naegleria fowleri* can infect the nasal mucosa and from there reach and proliferate in the brain. The fatality rate is nearly 100 percent.

Slow Virus Diseases

There are a number of fatal diseases of the human central nervous system caused by unconventional agents. They often are referred to as **slow virus diseases**, the name referring to the slow progress of these diseases. Sheep scrapie is a typical slow virus disease. Creutzfeldt-Jakob disease and Kuru are similar diseases in humans. The agents causing these diseases are unknown; they do not have any detectable nucleic acid, and one hypothesis is that they are comprised entirely of protein (prions). Chapter 12 has an interesting discussion of this subject.

Test

In the matching section, there is only one answer to each question; however, the lettered options (a, b, c, etc.) may be used more than once or, perhaps, not at all.

I. Match

1. A membrane layer covering the brain and spinal cord.

2. A slow virus disease.

3. Site of attack of the poliomyelitis virus.

4. Hansen's disease.

5. Human diploid cell vaccine.

6. Infective agent of protein with no detectable nucleic acid

a. Meninges

b. Anterior horn cells

c. Rabies

d. Kuru

e. Viroid

f. Prion

g. Leprosy

II. Match

1. Innermost layer of the meninges.

2. Outermost layer of the meninges.

3. Middle layer of the meninges.

a. Dura mater
b. Ventricles
c. Subarachnoid space
d. Arachnoid
e. Pia mater

III. Match

1. Treated by the Pasteur treatment.

2. Treated by human diploid cell vaccine.

3. Caused by a bullet-shaped rhabdovirus.

4. Also known by the term hydrophobia.

5. Thought to be transmitted by inhalation of the pathogen in dried pigeon droppings.

6. Caused by the bacterium *Neisseria meningitidis*.

7. Protozoan disease.

a. Rabies

b. Meningococcal meningitis

c. *Hemophilus influenzae* meningitis

d. *Cryptococcosis*

e. Poliomyelitis

f. Penumococcal meningitis

g. African trypanosomiasis

IV. Match

1. A slow virus disease.

2. Schwartzman phenomenon is an occasional characteristic.

3. The drugs of choice for treatment are amphotericin B or flucytosine.

4. Opposing muscles contract, causing spasms.

5. Pathogen grows at refrigerator temperatures.

a. Creutzfeldt-Jakob disease

b. Meningococcal meningitis

c. Listeriosis

d. *Crytococcus neoformans* meningitis

e. Tetanus

V. Match

1. Uses live viruses.

2. On rare occasions, the vaccine has caused the disease by a mutation to virulence.

a. Salk polio vaccine

b. Sabin polio vaccine

VI. Match

1. An amoebic protozoan found in ponds and streams; causes a lethal brain infection.

2. Spread by bite of tsetse fly.

3. An important cause of bacterial meningitis.

4. Cause of African sleeping sickness.

a. *Naegleria fowleri*

b. *Trypanosoma brucei gambiense*

c. *Cryptococcus neoformans*

d. *Streptococcus pneumoniae*

VII. Match

1. Probably most virulent; most common type in western United States.

2. Outbreaks often involve seafoods; nonproteolytic.

3. Toxin can be produced at refrigerator temperatures.

a. Type A botulism

b. Type B botulism

c. Type E botulism

VII. Fill-in-the-Blanks

1. An infection of the brain is called _____.

2. The _____ nervous system consists of the brain and spinal cord.

3. An infection of the meninges is called _____.

4. The _____ consists of capillaries that permit certain substances to pass from the blood to the brain but prevent others.

5. The _____ nervous system consists of all the nerves branching from the brain and the spinal cord.

6. The polio vaccine that uses viruses inactivated by formalin is called the _____ vaccine.

7. The bacterium causing _____ can be successfully grown in armadillos or on the footpads of mice.

8. Of the several types of arthropod-borne encephalitis that occur in the United States the most severe in its effects is _____.

9. The T in the DPT vaccine stands for _____.

10. *Hemophilus influenzae* is characterized, in part, by a growth requirement for _____ and _____ factors from lysed red blood cells.

11. For isolation of some pathogens such as *Neisseria meningitidis*, a _____ or carbon dioxide incubator is used to produce an environment of 5 to 10 percent carbon dioxide.

12. _____ fluid circulates in spaces within the brain (ventricles).

13. The most effective rabies vaccine, and the one with the fewest adverse effects such as allergic reactions, is the _____.

14. Nitrates that are used in some foods to prevent the growth of *Clostridium botulinum* may form _____ carcinogenic compounds.

15. Persons with a less effective cell-mediated immune system are more likely to develop the _____ form of leprosy.

Key

Matching

I. 1.a 2.d 3.b 4.g 5.c 6.f
II. 1.e 2.a 3.d
III. 1.a 2.a 3.a 4.a 5.d 6.b 7.g
IV. 1.a 2.b 3.d 4.e 5.c
V. 1.b 2.b
VI. 1.a 2.b 3.d 4.b
VII. 1.a 2.c 3.c

Fill-in-the-Blanks

1. encephalitis   2. central   3. meningitis   4. blood-brain barrier   5. peripheral
6. Salk   7. leprosy   8. Eastern equine encephalitis   9. tetanus   10. X and V
11. candle jar   12. cerebrospinal   13. human diploid cell vaccine   14. nitrosamines
15. lepromatous (progressive)

22 *Microbial Diseases of the Cardiovascular System*

The **cardiovascular system** consists of the heart, blood, and blood vessels. The **lymphatic system** consists of lymph, lymph vessels, lymph nodes, and lymphoid organs such as the thymus, spleen, and tonsils. The cardiovascular system can serve as a vehicle to spread infections in the body but are also, particularly the lymphatic system, the site of many of the body's defenses against infections.

Structure and Function of the Cardiovascular System

The **heart** is the center of the cardiovascular system (see Figure 22-1). **Blood** is pumped to the lungs where it exchanges carbon dioxide for oxygen and returns to the heart. This oxygenated blood is then pumped throughout the body. Circulating blood provides oxygen to tissue cells and picks up waste carbon dioxide. Blood passes from the heart in **arteries** and returns in **veins**. Arteries eventually branch into smaller vessels, **arterioles** (the equivalent in veins are **venules**), and then branch into even smaller vessels called **blood capillaries**. Capillary walls are only one cell thick. This thickness allows blood and tissue cells to exchange materials. The liquid of blood is **plasma**. Red blood cells, also called **erythrocytes**, carry oxygen. White blood cells, **leukocytes**, are important in defending against infection and include phagocytes and the B and T cells involved in acquired immunity. **Platelets** are involved in blood clotting.

Structure and Function of the Lymphatic System

Some plasma filters through blood capillary walls and, as **interstitial fluid**, bathes tissue cells. This fluid is picked up by **lymph capillaries** and is now referred to as **lymph**. Microorganisms also are able to enter lymph capillaries easily. Lymph eventually rejoins venous blood just before it reenters the heart. **Lymph nodes**, small bean-shaped structures, are located at various points along the lymphatic system. Phagocytic and antibody-producing cells help remove foreign bodies in the lymph or cause antibodies against them to be produced. The tonsils, appendix, spleen, and thymus gland also are lymphatic organs.

Bacterial Diseases of the Cardiovascular System

Septicemia

Uncontrolled proliferation of microbes in the blood is called **septicemia**. This condition causes fever and the drop in blood pressure known as **septic shock**. **Lymphagitis**, which becomes apparent as red streaks under the skin running up the arm or leg from an infection site, occurs when the lymph system becomes involved. Gram-negative rods are most commonly involved with septicemia today. The endotoxins released by the death and lysis of such cells are responsible for most of the symptoms. Gram-positive bacteria, such as the staphylococci, were a more common cause of such infections at one time.

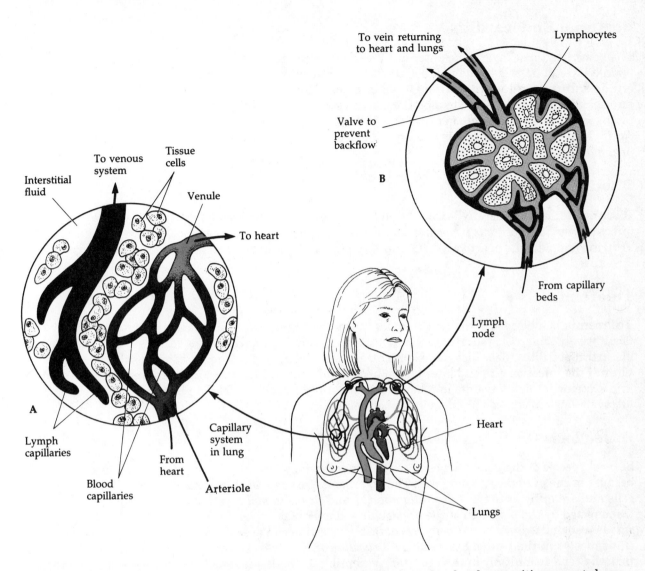

Figure 22-1. Details of the lung capillary system. Details of circulation to head and extremities are not shown in this diagram. Blood circulates from the heart through the arterial system to the blood capillaries in the lungs and other parts of the body. From these capillaries, the blood returns through the venous system to the heart. From the blood capillaries, some plasma filters into the surrounding tissue and enters the lymph capillaries. The inset shows the details of this exchange. This fluid, now called lymph, returns to the heart through the lymphatic circulatory system. All lymph returning to the heart must pass through at least one lymph node.

Puerperal Sepsis

An infection of the uterus that often leads to septicemia is called **puerperal sepsis**, also known as **puerperal fever** or **childbirth fever**. Hemolytic group A streptococci such as *Streptococcus pyogenes* are the most common cause. **Peritonitis**, an infection of the abdominal cavity, may be a final result. Semmelweiss and Holmes long ago clearly showed that poor hygienic practices by doctors and midwives were the primary cause, and that proper disinfection and sterilization could prevent transmission. Infections from abortions are mostly caused by anaerobic bacteria such as *Bacteroides* and *Clostridium* species.

Bacterial Endocarditis

The heart has three layers: the outer layer, called the **pericardium**, is a sac enclosing the heart; a middle layer of muscle tissue is called the **myocardium**; and the inner layer is called the **endocardium**. **Endocarditis** is an inflammation of the endocardium. **Subacute bacterial endocarditis** usually is due to alpha-hemolytic streptococci. Microbes from tonsillectomies and tooth extractions are a common cause of this disease. Blood clots formed are particularly dangerous. **Acute bacterial endocarditis** usually is caused by *Staphylococcus aureus* or *Streptococcus pneumoniae*.

Rheumatic Fever

Rheumatic fever generally is considered to be an autoimmune reaction by the body to repeated infections by *Streptococcus pyogenes*, usually from a streptococcal sore throat. Inflammations cause arthritis and damage heart valves. Group M proteins of the streptococci may be the main antigenic structure.

Tularemia

Tularemia is a disease of the lymph nodes caused by *Francisella tularensis*, a gram-negative, facultatively anaerobic rod. Hunters in contact with small animals such as rabbits are often infected, but arthropod bites (deer flies, for example), also may be involved. The infection results in a small ulcer at the infection site and later lymph node involvement. The microorganism requires the amino acid cysteine in high amounts in its medium and grows intracellularly in phagocytes. This intracellular location is a problem in chemotherapy.

Brucellosis (Undulant Fever)

Several species of the genus *Brucella* cause **brucellosis**, which also is known as **undulant fever** because of the periodic spiking fever. The organisms are capable of intracellular growth in phagocytic cells and are difficult to treat. *Brucella abortus* and *Brucella melitensis* are most commonly transmitted by the ingestion of the unpasteurized milk of cattle and goats, respectively. *Brucella suis* is transmitted most commonly by contact with swine carcasses and is the most common mode of transmission in the United States today. *Brucella canis* may be transmitted by contact with dogs. Brucella tend to multiply in the uterus of susceptible animals—favored by presence of mesoerythritol in the fetus and in surrounding membranes.

Anthrax

Robert Koch first identified **anthrax** as being caused by a spore-forming rod now called *Bacillus anthracis*. The spores are found in soil and ingested by animals while grazing. Humans can get anthrax by contact with spore-containing animal hides and other products. A malignant pustule forms at the site of entry on the skin and septicemia may result. Human gastrointestinal anthrax caused by ingestion of endospores has been reported. Woolsorter's disease is particularly dangerous; it is pneumonia caused by inhaling large numbers of the spores. The vaccine is based on the toxin produced by the organisms.

Gangrene

Anaerobic conditions occur in tissue if a wound interrupts the blood supply (**ischemia**). In turn, ischemia leads to death (**necrosis**) of tissue, or **gangrene**. Substances from dead cells provide nutrients for bacteria such as the endospore-forming anaerobes of the genus *Clostridium*. Gas gangrene, especially in muscle tissue, occurs when *Clostridium perfringens* ferments carbohydrates in wound tissue and produces carbon dioxide and hydrogen gases. This microorganism produces exotoxins and enzymes that further interfere with blood supply and otherwise favor the spread of the infection. This condition spreads the area of necrosis; eventually hemolytic anemia, and ultimately severe toxemia and death may ensue. Treatments for gas gangrene include debridement (surgical removal of tissue) or amputation. Hyperbaric chambers, in which the patient is subjected to an oxygen-rich atmosphere that interferes with the growth of the obligately anaerobic clostridia, are particularly useful for abdominal gangrene. Antibiotics such as penicillin and chloramphenicol also are used.

Systemic Diseases Caused By Animal Bites and Scratches

Animal bites inflicted on humans often become infected. *Pasteurella multocida*, which normally is more likely to cause septicemia in animals, can cause infections in humans. The infection may be localized at the site of the wound or may progress to pneumonia and life-threatening septicemia. Anaerobes such as *Clostridium*, *Bacteroides*, and *Fusobacterium* also can infect deep bites. Cat scratches can cause an ulcerated lesion and painful swelling of regional lymph nodes, a condition known as **cat scratch fever**. A minute gram-negative bacterium, which requires special staining to observe, is the suspected cause. Rat bites can cause recurring fever and arthritis-like symptoms known as **rat bite fever**.

Plague

In the Middle Ages **plague** was known as the Black Death because of the characteristic blackish areas on the skin caused by hemorrhages. The name **bubonic plague** derives from swellings, called buboes, that form as the lymph nodes in the groin and armpit regions become enlarged. The disease is caused by a gram-negative rod, *Yersinia pestis*, that is able to survive and even increase in number inside phagocytic cells. Normally the disease is transmitted by the rat flea from rat reservoirs, but in the western United States the disease is endemic in wild rodents (sylvatic plague). A particularly dangerous development is pneumonic plague, which arises when the bacteria are carried by the blood to the lungs. **Pneumonic plague**, because of airborne droplet infection, is highly infective. Untreated, plague has a mortality rate of 50 to 75 percent, and for pneumonic plague, the rate approaches 100 percent. Diagnosis is most commonly done by isolation and by identification of the isolated bacterium by fluorescent antibody tests or phage tests. Antibiotics such as streptomycin and tetracycline are useful for prophylactic protection of exposed persons and for treatment.

Relapsing Fever

All members of the genus *Borrelia* (a spirochete) cause relapsing fever. The disease is transmitted by ticks that feed on rodents and reptiles. A high fever is accompanied by jaundice and rose-colored spots. The fever subsides after about 10 days, but several relapses may occur. Diagnosis is made by observation of spirochetes in the patient's blood.

Lyme Disease

Lyme disease is caused by the bite of a tick. *Ixodes pacificus* on the Pacific coast and mostly *Ixodes dammini* in the rest of the United States. A skin lesion spreads from the site of the bite, clearing in the center. Later complications are arthritis and, occasionally, heart and neurological abnormalities. A spirochete, *Borrelia burgdorferi*, has been identified as the pathogen. A fluorescent antibody test will detect the pathogen in body fluids; it can be successfully treated by antibiotics.

Typhus

Several related diseases are caused by rickettsias that are spread by arthropod vectors such as mites, ticks, mosquitoes, fleas, and lice. **Epidemic typhus**, caused by *Rickettsia prowazekii*, is louse-borne. The pathogen is excreted in the louse feces and rubbed into the wound when the bitten host scratches the bite. Symptoms are a high and prolonged fever, stupor, and a rash of red spots caused by subcutaneous hemorrhaging. Mortality rates can be high but tetracycline and chloramphenicol are an effective treatment. Laboratory diagnosis is based on an indirect fluorescent antibody test and a positive agglutination test using latex beads. A vaccine is available. The related **endemic murine typhus** occurs sporadically rather than in epidemics. The most common hosts are rodents such as rats and squirrels. The pathogen, *Rickettsia typhi*, is transmitted by rat fleas. Diagnosis and treatment of both typhus diseases are similar.

Spotted Fever

The best known spotted fever in the United States is tick borne typhus, or **Rocky Mountain spotted fever**, which is most prevalent in the southeastern United States and Appalachia. The rickettsial pathogens are transmitted to humans by ticks. In the east, dog ticks mainly are responsible, and in the Rocky Mountains, wood ticks. The rickettsias usually are transmitted among ticks by transovarian passage, infecting tick eggs as they are produced. Clinically, the disease resembles the typhus fevers. Serological tests such as complement fixation are used in diagnosis, and a guinea pig inoculation, which produces a disease with typical symptoms, also may be used. Mortality rates now are less than 5 percent; chloramphenicol and tetracyclines are effective in treatment.

Viral Diseases of the Cardiovascular System

Myocarditis

Viruses, usually of the coxsackie (enterovirus) group may cause **myocarditis**. The virus reaches this layer of the heart via the blood or lymph and may cause scar tissue. The disease occurs in infants, where it is referred to as **myocarditis of the newborn**.

Infectious Mononucleosis

Mononucleosis is caused by a herpesvirus, the Epstein-Barr (EB) virus. The disease is characterized by enlarged and tender lymph nodes, enlarged spleen, fever, sore throat, headache, nausea, and general weakness. It probably is most commonly spread by saliva. As a result of the infection, mononuclear white blood cells proliferate in a manner resembling leukemia. It is rarely fatal. Diagnosis involves detection of nonspecific heterophil antibodies and by immunofluorescent techniques against the EB virus itself. Recovery results in a good immunity, but patients may continue to shed the virus because of latent infection. The EB virus also is suspected as a cause of **chronic fatigue**.

Burkitt's Lymphoma

The EB virus also causes two forms of cancer: Burkitt's lymphoma and nasopharyngeal cancer. Neither of these is very common in the United States. Fortunately, not many infected with the EB virus in this country develop these cancers.

Yellow Fever

Arboviruses (arthropod-borne viruses) reproduce in arthropods or in humans. **Yellow fever** is caused by the yellow fever virus of this type. Once injected by the mosquito vector, the virus multiplies in local lymph nodes; later the liver, spleen, kidneys, and heart are infected. Early stages of the disease are characterized by fever, chills, headache, nausea, and vomiting. This is followed by jaundice resulting from deposition of bile pigments in the skin and mucous membranes—a reflection of liver damage. Yellow fever is endemic in Central and tropical South America and Africa. Monkeys are a reservoir of the virus, but local control can be accomplished by localized control of the vector, *Aedes aegypti* mosquitoes, or by immunization. The vaccine is an attenuated live viral strain.

Dengue

Aedes aegypti also is the vector of **dengue**, milder than yellow fever and endemic in many tropical climates. Dengue is rarely fatal, but because of the muscle and joint pain experienced, the name of breakbone fever has been used. An Asian mosquito, *Aedes albopictus*, has been introduced into the United States. It transmits the virus by ovarian passage, and has a range that covers much of the country. For this reason, dengue may become more common. **Dengue hemorrhagic fever**, characterized by bleeding from the skin, gums, and gastrointestinal tract, may sometimes result in circulatory failure and shock.

Viral Hemorrhagic Fevers

A number of diseases are characterized by headache, muscle pain, skin or gastrointestinal hemorrhaging, and shock. Mortality rates often are high. Among them are **Lassa fever**, **Marburg virus**, **Crimean-Congo hemorrhagic fever**, and **Ebola hemorrhagic fever**.

Protozoan Diseases of the Cardiovascular System

Toxoplasmosis

The protozoan *Toxoplasma gondii*, which causes the disease of the blood and lymph vessels called **toxoplasmosis**, forms spores much like the malarial parasite. Cats are essential for reproduction of the organisms, as the sexual phase appears to occur only in their intestinal tract. Shed in cat feces, the protozoan infects a new host, usually by ingestion. Eating undercooked meats of an animal infected in this manner or inhalation of dried cat feces may cause human infections. In pregnant women, infection of the fetus across the placenta can cause drastic change to the fetus. A serological test using fluorescent antibody or hemagglutination testing, or both tests in conjunction, can be used in diagnosis.

American Trypanosomiasis (Chagas' Disease)

An example of a protozoan disease of the cardiovascular system is **American Trypanosomiasis (Chagas' disease)**. The causative agent is *Trypanosoma cruzi*, a flagellated protozoan. The disease occurs in the southern United States and throughout Mexico, Central America, and South America. The arthropod vector is the reduviid bug (Figure 11-29 in the text). Reservoirs for *T. cruzi* are various wild animals. The disease is transmitted to humans when insect bites are contaminated by the insect's feces. Most damage is caused by inflammatory reactions after transport by the blood to the liver, spleen, heart, etc. One symptom is loss of involuntary muscular contractions in the esophagus and gastrointestinal tract. These become grossly enlarged. Nifurtimox is the drug of choice but is not a cure.

Malaria

Malaria is characterized by chills and fever, and often vomiting and severe headaches. Symptoms typically occur in cycles of one to three days. Malaria, caused by the spore-forming protozoans of the genus *Plasmodium*, is found anywhere the protozoan parasite is present in human hosts, and the *Anopheles* mosquito is found. (It also can be spread by blood transfusions or contaminated syringe needles.) It is a serious and increasing problem in tropical Asia, Africa, Central America, and South America.

There are four forms of malaria: *Plasmodium falciparum* (most dangerous and geographically widespread), *Plasmodium vivax* (also widely distributed), *Plasmodium malariae*, and *Plasmodium ovale* (both of which have a lower infection rate and are geographically restricted, milder diseases). The infecting plasmodium, in the form of a sporozoite, enters liver cells where it undergoes schizogony. Later, merozoites are released into the bloodstream to infect red blood cells (some remain in the liver and may cause recurrences). When more merozoites are released from rupturing red blood cells, paroxysms of chills and fever result. Anemia from loss of red blood cells and hypertrophy of the liver and spleen are added complications. Some merozoites become male and female gametocytes, and are picked up by feeding mosquitoes. These then pass through a sexual cycle in the mosquito to produce sporozoites.

Laboratory diagnosis is made by identifying the protozoan in blood smears. Sickle-cell anemia victims tend to be resistant to malaria. Quinine derivatives such as primaquine and chloroquine are used in chemotherapy.

Schistosomiasis

Schistosomiasis is caused by a flatworm parasite, a fluke. The disease is a major world health problem. Waters become contaminated with ova excreted with human wastes. The motile larval form of *Schistosoma* is released from the ova and enters a species of snail. Eventually, the pathogen emerges from the snail as a fork-tailed **cercaria** carried by the bloodstream to the liver or urinary bladder (depending on the species of fluke) where they mature into an adult form, which produces a new supply of ova. Defensive reactions by the body to presence of ova cause tissue damage (granulomas). The disease causes damage such as abscesses and ulcers to the liver and to other organs such as the lungs or urinary system. Diagnosis is mainly by detection of flukes or ova in fecal or urine specimens.Praziquantel and oxamniquine are used in chemotherapy. **Swimmer's itch** sometimes troubles swimmers in lakes in the northern United States. This is an allergic reaction to cercaria of a similar parasite of wildfowl that does not mature in humans.

Test

In the matching section, there is only one answer to each question; however, the lettered options (a, b, c, etc.) may be used more than once or, perhaps, not at all.

I. Match

1. The largest vessels carrying oxygenated blood from the heart.

2. Walls are only one cell thick; aid exchange of materials.

3. Small veins that carry deoxygenated blood.

a. Capillaries
b. Venules
c. Arterioles
d. Veins
e. Arteries

II. Match

1. Clotting factor in blood.

2. Blood cells important in phagocytosis and antibody production.

3. A lymphoid organ.

4. Small, bean-shaped structures in lymphatic system; sites of considerable defensive activity by the body.

5. Fluid that bathes tissue cells after their passage through capillary walls.

6. Blood cells that carry oxygen.

a. Platelets
b. Erythrocytes
c. Leukocytes
d. Plasma
e. Lymph
f. Interstitial fluid
g. Lymph nodes
h. Tonsils

III. Match

1. Characterized by uncontrolled proliferation of bacteria in the blood.

2. Of autoimmune origin due to group M proteins of streptococci.

3. Inflammation of the inner layer of the heart linings.

4. A common cause is microorganisms from tonsillectomies and tooth extractions.

5. The middle layer surrounding the heart; mostly muscle tissue.

6. Characterized by red streaks from the site of infection.

a. Lymphagitis
b. Septicemia
c. Pericardium
d. Endocarditis
e. Myocardium
f. Rheumatic fever
g. Myocarditis

IV. Match

1. Probably transmitted by saliva.

2. Childbirth fever; a form of septicemia.

3. Often transmitted by contact with small animals such as rabbits.

4. Undulant fever, at one time transmitted by ingestion of contaminated milk, is now transmitted by contact with animal carcasses.

5. Caused by a spore-forming rod that is present in the soil.

6. The cat is essential in the reproductive cycle and the transmission of the causative organisms.

7. Caused by a protozoan that forms spores.

8. Heterophil antibodies used in diagnosis.

9. Caused by the Epstein-Barr virus.

a. Brucellosis
b. Toxoplasmosis
c. Anthrax
d. Rocky Mountain spotted fever
e. Tularemia
f. Peurperal sepsis
g. Infectious mononucleosis

V. Match

1. Bite of a tick transmits a spirochete, *Borrelia burgdorferi*.

2. A swimming organism called a cercaria is an essential part of the life cycle of the pathogen.

3. A rickettsial disease transmitted by dog ticks or wood ticks.

4. A rickettsial disease transmitted by a louse.

5. A rickettsial disease transmitted by a rat flea.

6. Chagas' disease.

7. A spore-forming protozoan is the cause.

8. A viral hemorrhagic fever.

a. Gangrene
b. Lyme disease
c. Epidemic typhus
d. Rocky Mountain spotted fever
e. Endemic murine typhus
f. Yellow fever
g. Plague
h. Lassa fever
i. American trypanosomiasis
j. Malaria
k. Schistosomiasis

VI. Fill-in-the-Blanks

1. The surgical removal of tissue is called _____.

2. The fluid portion of the blood is called _____.

3. A general name for a white blood cell is _____.

4. When a gram-negative bacterium lyses, it releases harmful _____.

5. The outer sac enclosing the heart is called the _____.

6. Acute bacterial endocarditis usually is caused by _____ and _____ bacteria (give genus and species).

7. Group M proteins are associated with what genus of bacteria? _____

8. Infection of the abdominal cavity is called _____.

9. A bacterial species classified as hemolytic group A streptococcus is _____.

10. *Francisella tularensis* causes _____ .

11. *Brucellis suis* is found in what type of animals? _____.

12. A characteristic of _____ is malignant pustule that forms at the site of entry through the skin.

13. Coxsackieviruses are a common cause of a heart disease called _____.

14. Burkitt's lymphoma is caused by the same virus that causes _____.

15. Infections caused by obligate anaerobes such as *Clostridium perfringens* are sometimes treated by putting the patients in _____ chambers.

16. The causative organism of _____ has an unusual requirement for the amino acid cysteine and grows intracellularly in phagocytes.

17. Abortions are likely to cause infections by anaerobic bacteria of the genus _____ and _____.

18. A disease known to be transmitted by arthropods such as deer flies is _____.

19. Many years ago, Semmelweiss showed how proper hygiene and disinfection of hands and instruments could prevent _____.

20. Two diseases in which the microbial cell grows inside the host cell and is therefore relatively resistant to immune reactions and chemotherapy are _____ and _____.

21. When malaria is transmitted by the bite of a mosquito, the parasite form injected is the _____.

22. The most dangerous form of malaria is caused by *Plasmodium* _____.

23. Snails are essential to the life cycle of the disease organism causing _____.

24. The bite of the *Anopheles* mosquito causes _____.

Key

Matching

I. 1.e 2.a 3.b
II. 1.a 2.c 3.h 4.g 5.f 6.b
III. 1.b 2.f 3.d 4.d 5.e 6.a
IV. 1.g 2.f 3.e 4.a 5.c 6.b 7.b 8.g 9.g
V. 1.b 2.k 3.d 4.c 5.e 6.i 7.j 8.h

Fill-in-the-Blanks

1. debridement   2. plasma   3. leukocyte   4. endotoxins   5. pericardium
6. *Staphylococcus aureus*; *Streptococcus pneumoniae*   7. *Streptococcus*   8. peritonitis
9. *Streptococcus pyogenes*   10. tularemia   11. swine   12. anthrax   13. myocarditis
14. infectious mononucleosis   15. hyperbaric   16. tularemia
17. *Bacteroides*; *Clostridium*   18. tularemia   19. puerperal sepsis
20. tularemia; brucellosis   21. sporozoite   22. *falciparum*   23. schistosomiasis
24. malaria

23 *Microbial Diseases of the Respiratory System*

Structure and Function of the Respiratory System

The **upper respiratory system** consists of the nose and throat, including the middle ear and auditory tubes (Figure 23-1). Mucus traps dust and microorganisms and cilia assist in moving these to the throat for elimination. The **lower respiratory system** consists of the larynx, trachea, bronchial tubes, and lungs (Figure 23-2). **Alveoli** are air sacs in lung tissue where oxygen and carbon dioxide are exchanged, and the double-layered membrane around the lungs is the **pleura**. Macrophages in the alveoli destroy many pathogens, and IgA antibodies in mucus, saliva, and tears aid in resistance.

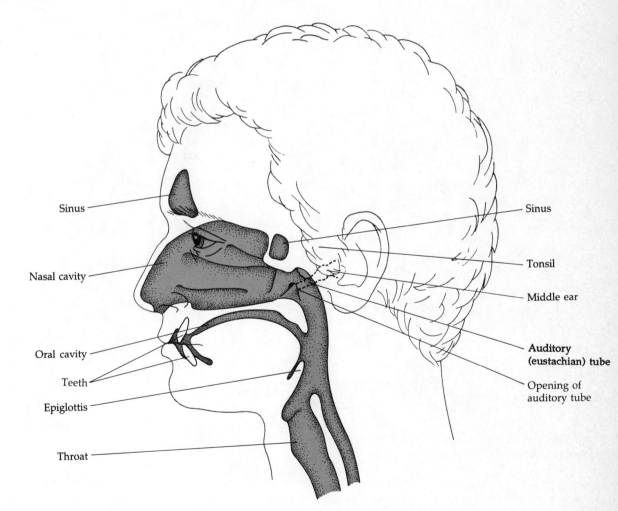

Figure 23-1. Structures of the upper respiratory system.

Normal Flora of the Respiratory System

Nasal cavity flora includes diphtheroids, staphylococci, micrococci, and *Bacillus* species—all gram-positive bacteria. The throat near the nose may contain streptococci, *Hemophilus influenzae*, and *Neisseria meningitidis*. The trachea contains few bacteria, and the lower respiratory tract usually is sterile.

Diseases of the Upper Respiratory System

Among the diseases of the upper respiratory system are **pharyngitis**, a sore throat; **laryngitis**, an infected larynx (see Figure 23-1); **tonsilitis**, inflamed tonsils; and **sinusitis**, an infected sinus. **Epiglottitis**, an inflammation of a flap-like structure of cartilage that prevents material from entering the larynx, is a possibly life-threatening disease. It usually is caused by *Hemophilus influenzae* type b.

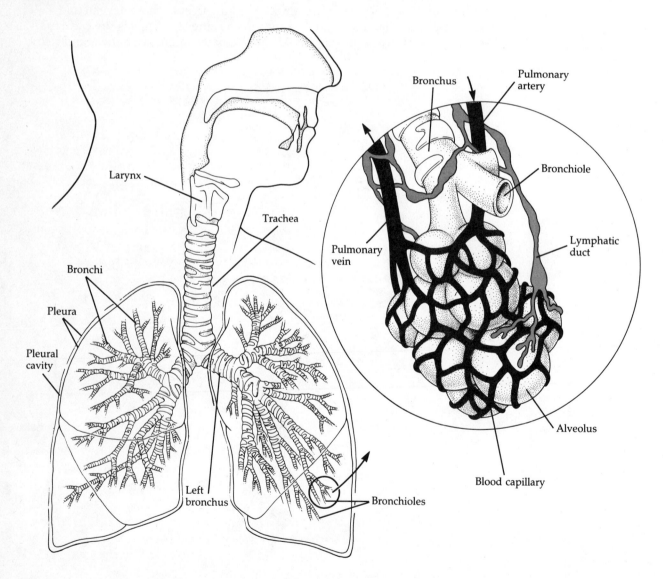

Figure 23-2. Structures of the lower respiratory system.

206

Bacterial Diseases of the Upper Respiratory System

Streptococcal Pharyngitis (Strep Throat)

Streptococcus pyogenes, a beta-hemolytic group A streptococcus, is an important cause of **streptococcal pharyngitis**. Diagnosis makes use of new latex particle passive agglutination text which may require as little as ten minutes to perform. Treatment usually is with penicillin or erythromycin. Symptoms are inflammation and fever; complications are tonsillitis or otitis media.

Scarlet Fever

Some strains of *S. pyogenes* have been lysogenized by a phage and produce an erythrogenic toxin that causes **scarlet fever**. The skin affected by the toxin turns pink-red, and a high fever occurs. Affected skin may peel, and deafness is an occasional complication. Immunity is developed only to that serological strain of the streptococcus that causes the disease.

Diphtheria

Corynebacterium diphtheriae, a gram-positive, non-spore-forming pleomorphic rod, causes **diphtheria**. Dividing cells often fold into V- and Y-shaped figures. Many well persons are symptomless carriers. **DPT vaccine** (D for diphtheria) is a toxoid that stimulates immunity to the diphtheria toxin. Immunity is measured by the Schick test. After the toxin is injected, a lack of reaction shows that the subject has neutralizing antibodies and is immune. Exotoxin, whose production is controlled by a gene acquired from a lysogenic phage, is carried by the bloodstream and interferes with protein synthesis in the host's cells. A membrane composed of fibrin and dead human and bacterial cells (diphtheria meaning "leather") may form in the throat and block air passage.

Laboratory diagnosis can be made by bacterial isolation on Loeffler agar slants and tellurite agar. Typical colonies are formed on these media, and microbes can be isolated for guinea pig virulence and gel diffusion testing. Antitoxins are essential in treatment but must be administered before toxin enters the cells. Antibodies also are useful but are not a substitute for antitoxin. **Cutaneous diphtheria** is responsible for many cases in older persons who have no or weakened immunity.

Otitis Media

Infections of the upper respiratory system may cause **otitis media**, or inner ear infection leading to earache. The symptoms are due to pus formation and pressure against the eardrum. Enlarged adenoids and the relatively small auditory tubes in many children contribute to susceptibility. *Staphylococcus aureus*, *Streptococcus pneumoniae*, *Hemophilus influenzae*, and *Branhamella catarrhalis*, are the most commonly identified causes of this disease. Broad spectrum penicillins are the first choice in chemotherapy.

Viral Disease of the Upper Respiratory System

Common Cold (Acute Coryza)

The **common cold** or **acute coryza** is caused mainly by viruses classified as coronaviruses or rhinoviruses. There are more than 200 viruses that cause colds, so immunity to any one is of limited use. Immunity develops in isolated populations due to IgA antibodies, and immunities accumulate with age. Laryngitis and otitis media are common complications. Symptoms can be alleviated but

no cure exists.
Diseases of the Lower Respiratory System

A lower respiratory tract infection affecting the bronchi (Figure 23-2) is **bronchitis** or **bronchiolitis**. A severe complication of bronchitis is **pneumonia**. These interrelated diseases are sometimes termed **croup**.

Whooping Cough (Pertussis)

Infection by *Bordetella pertussis*, a gram-negative coccobacillus, can cause **whooping cough** (**pertussis**). Ciliary action is blocked by accumulations of dense masses of these bacteria in the trachea and bronchi. Eventually, *tracheal cytotoxin* causes loss of ciliated cells. The initial, **catarrhal**, stage resembles a common cold. The **paroxysmal** stage is characterized by attempts to clear accumulations by coughing. The bacteria may be cultured on Bordet-Gengou medium, where typical colonies are formed. Fluorescent antibody tests may confirm identification. Erythromycin, chloramphenicol, and tetracyclines are useful antibiotics. Recovery results in good immunity. Vaccination is recommended although there is some concern about adverse effects. (The P in DTP vaccine stands for pertussis.)

Tuberculosis

Mycobacterium tuberculosis, a slow-growing rod that sometimes forms filaments, is the cause of **tuberculosis**. The organism is *acid-fast*, meaning it retains the carbol fuchsin dye of the Ziehl-Neelson staining technique. The stain characteristically reflects the large lipid content of the cell wall, which also may make it resistant to drying, sunlight, and chemical disinfectants. Tubercle bacilli entering the lungs are phagocytized by macrophages in the alveoli (Figure 23-2). The pathogens often continue to proliferate in the macrophages. Infection results in nodule formation, usually in the lungs after inhalation of the pathogen. Transmission also can be by ingestion or contact. A hypersensitivity reaction against tubercle bacilli released from dying macrophages results in formation of **tubercles**. In time these may undergo necrosis, changing to caseous lesions which may progress to form a tuberculosis cavity. When healed, these lesions show in x-rays as Ghon complexes. Attempts to wall off the pathogen in tubercles may fail, and the infection may spread to new sites, resulting in a condition called **miliary tuberculosis**. In this form, many tubercles form and a progressive disease ensues.

Stress and genetic differences contribute to susceptibility to tuberculosis. The intracellular growth of tuberculosis shields from antibiotics. Drugs such as streptomycin, rifampin, isoniazid (INH), and ethambutol are used as treatment, usually in combinations to aovid resistance. Skin tests such as the Mantoux test involve injecting a purified protein derivative (PPD) of the tuberculosis bacterium. T cells sensitized by exposure to tuberculosis cause a delayed hypersensitivity reaction at the injection site. *Mycobacterium bovis* is the cause of bovine tuberculosis, which usually is spread by unpasteurized milk. This disease mainly affects bones and the lymphatic system. A vaccine, BCG, is a live attenuated culture of *M. bovis*. Tuberculosis, which the body combats by cell-mediated immunity, is a frequent complication of AIDS.

Bacterial Pneumonias

The term pneumonia is applied to many pulmonary infections.

Pneumococcal pneumonia. **Pneumococcal pneumonia**, caused by the pneumococcus *Streptococcus pneumoniae*, is the classic form of pneumonia. The organism can be distinguished from other similar organisms by inhibition of its growth near a disk of optochin, or by bile solubility. Serological typing using the quellung reaction is useful. A vaccine is used to immunize elderly or debilitated individuals. Penicillin is the drug of choice in treatment.

Klebsiella pneumonia. Chronically debilitated persons, alcoholics for example, may get pneumonia from *Klebsiella pneumoniae*, a common throat organism and opportunistic pathogen. Lung abscesses and permanent lung damage may occur, and there is a high fatality rate.

Mycoplasmal pneumonia (primary atypical pneumonia). Pneumonias typically have a bacterial etiology. If a bacterial agent is not isolated, the pneumonia is considered atypical. The most common atypical type (**primary atypical pneumonia**) is caused by a bacterium with no cell wall, *Mycoplasma pneumoniae*, which does not grow on routine isolation media. The bacterium can, however, be grown and isolated on horse serum and yeast extract medium. The disease is relatively mild and sometimes called **walking pneumonia**. Tetracyclines often are effective in treatment. Diagnosis may use a complement fixation test to detect a rise in titer. A DNA-probe test has recently been developed.

Legionellosis. Legionellosis (**Legionnaires' disease**) is caused by a gram-negative rod, *Legionella pneumophila*. It may grow in waters of air-conditioning cooling towers or in the hot water lines of hospitals. Erythromycin is the drug of choice for treatment.

Psittacosis (Ornithosis). Psittacine birds such as parakeets and other parrots carry this disease, hence the name **psittacosis**, or **ornithosis**. *Chlamydia psittaci*, a gram-negative obligately intracellular bacterium, is the cause. It forms **elementary bodies**, which are the airborne infective agent. Within the cell they develop into larger **reticulate bodies** that reproduce by fission; eventually they become elementary bodies and leave the cell. The dried droppings of sick birds are the usual source of these. Diagnosis is made by isolation in embryonated eggs, mice, or cell culture. Fluorescent antibody testing is used for final identification of isolated organisms, and a complement fixation test is used to detect serum antibodies.

Q fever. *Coxiella burnetii*, an obligately intracellular rickettsia, causes **Q fever**. It is a parasite of cattle ticks that eventually contaminates dairy products. Diagnosis is made by isolation of the organism in egg embryo or cell culture.

Viral Diseases of the Lower Respiratory System

Viral Pneumonia

Viral pneumonia is seldom confirmed by culture, but is often assumed if mycoplasmal and bacterial pneumonia are ruled out. **Respiratory syncytial virus** is the most common cause—especially in children.

Influenza (Flu)

The **influenza** virus has an envelope with two types of projections: *hemagglutinin* (H) *spikes* and *neuraminidase* (N) *spikes*. Antibodies against influenza are directed mainly at these spikes. Variation in their composition alters the antigenic type of the virus and helps it evade established resistance in the human population. *Type A* of the virus is more widespread and severe, and *type B* is milder and geographically limited. *Antigenic drift* is the result of minor variations of antigenic makeup; *antigenic shifts* are major changes. Vaccines are available to high risk groups; they are usually multivalent, directed at the types then in circulation. These latter cause the designation — the Hong Kong virus of 1968, for example—to change and cause the virus to evade almost all established immunity. Amantadine, if taken very early, will significantly reduce symptoms.

Fungal Diseases of the Lower Respiratory System

Histoplasmosis

Histoplasmosis is caused by the dimorphic fungus, *Histoplasma capsulatum*, and may resemble tuberculosis. It is, however, much milder. It is found mainly in the Mississippi and Ohio River valleys. Bird droppings in the soil encourage growth of the fungus, which is spread by airborne conidia. Amphotericin B or ketoconazole are used in chemotherapy.

Coccidioidomycosis

A dimorphic fungus, *Coccidioides immitis*, whose sources are found in the soil of the American Southwest, is the cause of **coccidioidomycosis**. A tuberculinlike skin test is used in screening. Amphotericin B and ketoconazole are used to treat the disease, but most cases are subclinical; about 1 percent are progressive—resembling tuberculosis.

Blastomycosis (North American Blastomycosis)

Another dimorphic fungus, *Blastomyces dermatitidis*, causes North American **blastomycosis**. It is found most often in the Mississippi Valley. Although most cases are subclinical, a rapidly progressive infection with cutaneous ulcers and abscess formation can occur. Amphotericin B is used in treatment.

Other Fungi Involved in Respiratory Disease

Other fungi such as *Aspergillus fumigatis* and other species of *Aspergillus* may cause **aspergillosis**. Compost piles are likely growth sites. Other mold genera also may cause pulmonary infections in susceptible individuals. **Nocardiosis**, caused by a soil actinomycete, usually is mild but occasionally is serious.

Protozoan Disease of the Lower Respiratory System

Pneumocystis pneumonia is caused by the sporozoan *Pneumocystis carinii*. The parasite may be transmitted by direct human contact, mostly to immunosuppressed persons. Pneumocystis causes the alveoli to become filled with a frothy exudate; untreated infections usually are fatal. This disease is a common condition among the immunosuppressed—especially AIDS patients. Pentamidine is the drug of choice.

Test

In the matching section, there is only one answer to each question; however, the lettered options (a, b, c, etc.) may be used more than once or, perhaps, not at all.

I. Match

1. Caused by a lysogenized bacterium.

2. Pertussis.

3. The DTP vaccine helps prevent this.

4. Infection of the inner ear.

5. Ghon complexes are one possible characteristic .

6. The Mantoux test is useful in screening for this disease.

7. Caused by an acid-fast organism.

8. Ethambutol, isoniazid, and rifampin are useful in treatment.

9. A disease of the respiratory tract near the larynx caused by *Hemophilus influenzae*.

10. The causative bacterium may be isolated and identified on Bordet-Gengou medium.

a. Scarlet fever

b. Otitis media

c. Whooping cough

d. Tuberculosis

e. Acute epiglottitis

II. Match

1. Infectious agent causing the disease is an elementary body formed intracellularly by a bacterial pathogen.

2. Caused by *Coxiella burnetii*.

3. Caused by dimorphic fungus growing in bird droppings.

4. Caused by a dimorphic fungus widely distributed in airborne spores in the American Southwest.

5. Caused by rickettsial organisms parasitic of cattle ticks.

6. A fungal disease found with considerable frequency in the Ohio-Mississippi River valleys.

7. Acute coryza.

8. Also called ornithosis.

9. Caused mainly by viruses classified as coronaviruses or rhinoviruses.

a. Common cold

b. Histoplasmosis

c. Coccidioidomycosis

d. Q fever

e. Psittacosis

III. Match

1. Mycoplasmal pneumonia.

2. The causative organism is suspected to grow in water cooling towers of air-conditioning systems.

3. A pneumonia often affecting debilitated, poorly nourished individuals such as alcoholics.

4. Production of the harmful exotoxin involved in this disease is controlled by a lysogenic phage.

5. Pharyngitis.

6. DPT vaccine is useful in prevention.

a. Diphtheria

b. Legionellosis

c. Primary atypical pneumonia

d. Klebsiella pneumonia

e. Streptococcal sore throat

IV. Fill-in-the-Blanks

1. The designations H and N applies to what disease?_____.

2. The air sacs in lung tissue where oxygen and carbon dioxide are exchanged are the _____.

3. The double-layered membrane around the lungs is the _____.

4. Among diseases of the upper respiratory tract are pharyngitis, laryngitis, tonsilitis, sinusitus, and epiglottitis. Of these, only _____ is likely to be life-threatening.

5. The Schick test is used to measure immunity to _____.

6. Immunity to the common cold is based mainly on antibodies of the _____ type.

7. Tuberculosis spreading progressively to new sites is termed _____.

8. The vaccine used to help prevent tuberculosis is called _____.

9. The quellung reaction is used in diagnosis or serological typing of what disease? _____.

10. Older persons with no, or weakened, immunity often come down with _____ diphtheria.

11. The Ziehl-Neelson staining technique is valuable in diagnosis of _____.

12. Tuberculosis bacteria entering the lungs may be ingested by alveolar _____.

13. Minor year-to-year variations in the antigenic makeup of the influenza virus is called antigenic _____.

14. Major changes in the antigenic makeup of the influenza virus that allows it to evade almost all previous immunity is called antigenic _____.

15. The trachea contains few bacteria and the lower respiratory tract is ususally _____.

16. Laboratory diagnosis of _____ can be made by bacterial isolation of Loeffler agar slants and tellurite agar.

17. The terms catarrhal stage and paroxysmal stage refer to the disease _____.

18. The influenza virus has an envelope with two types of projections: _____ spikes and _____ spikes.

19. Diagnosis for streptococcal sore throat involves streaking a throat swab of bacteria onto blood agar to observe the type of _____.

20. A disease for which administration of antitoxin is essential in treatment is _____.

21. A protozoan-caused pneumonia often seen in AIDS patients is _____.

22. The rickettsial organism, *Coxiella burnetii*, is the cause of the disease _____.

Key

Matching

I. 1.a 2.c 3.c 4.b 5.d 6.d 7.d 8.d 9.e 10.c
II. 1.e 2.d 3.b 4.c 5.d 6.b 7.a 8.e 9.a
III. 1.c 2.b 3.d 4.a 5.e 6.a

Fill-in-the-Blanks

1. influenza 2. alveoli 3. pleura 4. epiglottitis 5. diphtheria 6. IgA
7. miliary 8. BCG 9. Pneumococcal pneumonia 10. cutaneous 11. tuberculosis
12. macrophages 13. drift 14. shift 15. sterile 16. diphtheria
17. whooping cough 18. hemagglutinin, neuraminidase 19. hemolysis 20. diphtheria
21. Pneumocystis 22. Q fever

24 *Microbial Diseases of the Digestive System*

Structure and Function of the Digestive System

The first group of organs of the essentially tubelike **gastrointestinal (GI) tract**, or **alimentary canal**, is the mouth, pharynx (throat), esophagus (food tube), stomach, small intestine, and large intestine (see Figure 24-1). Accessory structures consist of the teeth, tongue, salivary glands, liver, gallbladder, and pancreas. Except for the teeth and tongue, these lie outside the GI tract and produce secretions released into the tract. The action of secretions such as bile, pancreatic enzymes, saliva, and stomach and intestinal enzymes converts ingested foods into their end products of digestion. By the time the liquefied food leaves the small intestine, the absorption of sugars, fatty acids, and amino acids produced by digestion is almost complete. Water, vitamins, and nutrients are absorbed in the large intestine.

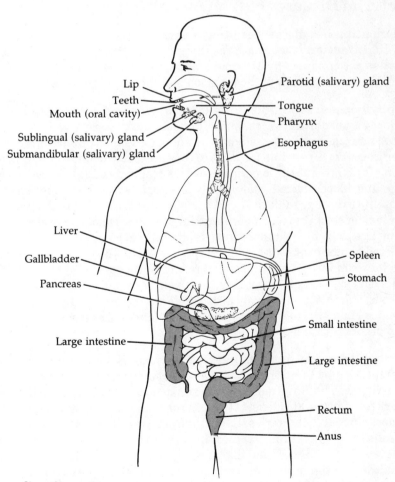

Figure 24-1. Anatomy of the human digestive system.

Normal Flora of the Digestive System

Saliva contains millions of bacteria per milliliter; several species of *Streptococcus* colonize the tongue, mucosa of the cheek, and the teeth. *Bacteroides*, *Fusobacterium* (anaerobes), and spirochetes inhabit spaces where gums and teeth merge. The stomach and small intestine have few microorganisms, but the large intestine contains as many as 100 billion per gram of feces. Most are anaerobes or facultative anaerobes, and some synthesize vitamins useful to the host.

Bacterial Diseases of the Mouth

Dental caries (tooth decay). The chief bacterial species responsible for tooth **caries** is *Streptococcus mutans*. It adheres to the teeth, producing **dextran** from the glucose component of sucrose, and forming **dental plaque**. Localized in the plaque, it produces acid from available carbohydrates, particularly fructose released from sucrose, and forms lactic acid, which leads to softening of the enamel and dentin of the tooth.

Peridontal disease. Peridontal disease is a collective term for a number of conditions characterized by inflammation and degeneration of the gums, supporting bone, etc. *Actinomyces*, *Nocardia*, and *Corynebacterium* species growing in the gingival crevice probably initiate the disease by forming the initial plaque leading to the inflammation.

Bacterial Diseases of the Lower Digestive System

Intoxication is a disease resulting from the ingestion of toxins preformed in food by microbial growth. **Infection** is a disease resulting from microbial growth in the tissues of the host. Severe diarrhea accompanied by blood and mucus is termed **dysentery**. The general term **gastroenteritis** describes inflammation of the stomach and intestinal mucosa.

Staphylococcal food poisoning (Staphylococcal Enterotoxicosis). *Staphylococcus aureus* produces an exotoxin (which, since it affects the digestive system, is called an **enterotoxin**) that causes nausea, vomiting, and diarrhea about one to six hours after ingestion. Serological type A enterotoxin is responsible for most of these cases. The enterotoxin is heat-resistant and withstands boiling for 30 minutes. Coagulase-positive staphylococci produce **coagulase**, an enzyme that coagulates blood plasma, and usually is associated with production of the toxin. Staphylococci are resistant to drying, radiation, and osmotic pressures. Custard, cream pies, and ham are foods with a high incidence of this food poisoning, largely because of the presence of sugar concentrations or curing agents, which suppress nonstaphylococcal competition.

Food can be contaminated by staphylococci on fingers or in skin lesions. If allowed incubation time for toxin formation, illness can result. Pathogenic staphylococci ferment mannitol, produce hemolysis, are coagulase-positive, form yellow colonies, and cause no obvious spoilage when growing on foods. There is no routinely practicable way to test for the presence of enterotoxin in foods. The presence of an enzyme, **thermostable nuclease**, is easily detected and an indicator of probable toxin production in a food.

Salmonellosis (Salmonella gastroenteritis). Salmonella are not assigned to species, but are differentiated by hundreds of serovars. All serovars of *Salmonella* are considered pathogenic to some degree. The Kauffmann-White serological scheme to identify serovars assigns numbers and letters to different antigens: O (somatic or body), Vi (capsular), and H (flagellar). This scheme is used by most clinical laboratories. **Salmonellosis** results from ingestion of the organisms and growth of the cells in the intestinal tract. An incubation time of about 12 to 26 hours is normal. Symptoms are moderate fever, nausea, and diarrhea; there is a low mortality rate. Salmonellae are common inhabitants of the intestinal tract of many animals; poultry and egg products are particularly common sources of food contamination.

Typhoid fever. *Salmonella typhi* is a very virulent serovar of *Salmonella* and the cause of typhoid fever. Incubation is about two weeks. Typical symptoms are fever and malaise. Diarrhea appears only late in the disease—the second or third week. In severe cases, there may be perforation of the intestinal wall and dissemination of the organism in the body. An occasional characteristic of the disease is **carriers**, who continue to spread the disease in feces after recovery. Antimicrobials such as chloramphenicol, ampicillin, and trimethoprim-sulamethoxazole are effective treatments.

Bacillary Dysentery (Shigellosis). Bacteria of the genus *Shigella*—*S. sonnei, S. dysenteriae, S. flexneri*, and *S. boydii*—all cause **bacillary dysentery** (**shigellosis**). *S. sonnei*, the most common type in the United States, causes a relatively mild disease, while *S. dysenteriae*, which is not common in the United States, cause a much more severe disease.

Cholera. *Vibrio cholerae*, the cause of **cholera**, is spread by the fecal-oral route. The incubation period is about three days. The organism produces an enterotoxin in the intestines that causes rapid loss of body fluids and electrolytes, as well as violent vomiting. The serogroup 0:1, biotype eltor, of *V. cholerae* causes the epidemic form of the disease. *Vibrio cholerae* not belonging to serogroup 0:1 (non-0:1) cause a similar disease that does not have the epidemic characteristics of cholera. These organisms are indigenous in much of Pacific and Gulf Coastal waters. They are more likely to be invasive and cause bloody stools and fever. Treatment is based mainly on replacement of lost fluids and electrolytes; antibiotics are not very effective. Untreated, the death rate can be as high as 50 percent; it is about 1 percent when properly treated.

***Vibrio parahaemolyticus* Gastroenteritis.** *Vibrio parahaemolyticus* causes a gastroenteritis similar to Asiatic cholera but milder. Incubation time is about 24 hours. Because it is present in coastal waters, outbreaks are usually associated with seafood. It is the most common gastroenteritis in Japan. The organism is halophilic and growth media must contain 2 to 4 percent sodium chloride.

***Escherichia coli* Gastroenteritis (Traveler's Diarrhea).** One of the most numerous microorganisms in the human intestinal tract, *Escherichia coli* normally is harmless, but some strains can cause gastroenteritis. Pathogenic strains possess adhesins such as pili to attach to intestinal mucosa. These strains may produce an enterotoxin (**enterotoxigenic** *E. coli*), invade the lining of the large intestine (**enteroinvasive** *E. coli*), or have no well-understood mechanisms (**enteropathogenic** *E. coli*). The organisms may be the major cause of traveler's diarrhea, and may cause outbreaks of epidemic diarrhea in nurseries.

***Campylobacter* Gastroenteritis.** *Campylobacter* are gram-negative, spirally curved rods that are part of the intestinal flora of animals such as sheep and cattle. Recently they have been identified as being one of the most common causes of diarrheal illness in the United States. ***Campylobacter* gastroenteritis** is characterized by fever, cramping abdominal pain, and diarrhea. There is speculation that *Campylobacter pylori* may be the cause of gastric ulcers.

***Yersinia* Gastroenteritis.** The symptoms of **yersiniosis** are diarrhea, fever, headache and abdominal pain severe enough to cause frequent misdiagnosis as appendicitis. It is caused by *Yersinia enterocolitica* and *Y. pseudotuberculosis*, gram-negative inhabitants of the intestines of many domestic animals. Both organisms are capable of growth at refrigerator temperatures.

***Clostridium perfringens* Gastroenteritis.** A very common, if often unrecognized, form of food poisoning in the United States is diarrhea caused by *Clostridium perfringens*. This endospore-forming anaerobe is the same organism responsible for gas gangrene; it produces a number of damaging toxins. Cooked meats or stews that contain meats are a common source of the illness. Cooking lowers the oxygen level for clostridial growth, and the spores survive the heating. An exotoxin is produced as the organism grows in the intestinal tract, causing abdominal pain and diarrhea. Most cases are not clinically diagnosed. Symptoms appear in 8 to 12 hours.

Bacillus cereus Gastroenteritis. *Bacillus cereus* is a gram-positive, spore-forming bacterium that produces toxins causing either a diarrheal episode resembling a *C. perfringens* infection, or nausea and vomiting.

Viral Diseases of the Digestive System

Mumps

The **parotid glands**, just below and in front of the ear, are the particular target of the mumps virus, a paramyxovirus. **Mumps** is characterized by inflammation and swelling of the parotid glands accompanied by fever and pain. Orchitis, an inflammation of the testes, also may occur. Incubation is about 16 to 18 days. Introduction of the measles-mumps-rubella (MMR) vaccine has caused a sharp decline in cases in recent years.

Cytomegalovirus (CMV) Inclusion Disease

Viruses of the herpesvirus family that produce a cellular response characterized by **cytomegaly** (large size) and the presence of intranuclear inclusion bodies are referred to as **cytomegalovirus (CMV)** diseases. Pregnancy and other stresses may activate latent CMV infections, which frequently affect the salivary glands. A mother may transmit her latent infection of CMV by congenital infection of the fetus by placental transfer. **Congenital cytomegalovirus inclusion** disease is a severe disease in newborns. Older victims commonly include cancer patients, those receiving immunosuppressive therapy, and patients receiving large volumes of blood. Stillborns show CMV lesions; antibodies in the blood of the general population are not uncommon.

Hepatitis

Hepatitis A (Infectious Hepatitis)

Hepatitis A, or **infectious hepatitis**, is a disease primarily affecting the liver. It is caused by the hepatitis A virus (**HAV**). It is primarily transmitted by the fecal-oral route. The virus is shed in the feces. Symptoms are loss of appetite (*anorexia*), malaise, nausea, diarrhea, abdominal discomfort, fever, chills lasting for two days to three weeks. *Jaundice*, a skin yellowing caused by liver infection, eventually appears. Mortality is less than 1 percent, and the disease often is subclinical. The incubation period is one to six weeks, which complicates epidemiology.

Hepatitis B (Serum Hepatitis)

Hepatitis B, or **serum hepatitis**, is caused by the hepatitis B virus (**HBV**). This is a completely different virus than HAV, as can be seen in Table 24-1, which summarizes several different types of hepatitis. HBV is most commonly transmitted by blood, but can be transmitted by many body fluids, such as saliva, breast milk, and semen. Three distinct structures with the hepatitis B surface antigen (HB_sAg) can be found in patients' blood. **Dane particles** probably represent intact hepatitis B virions, while **filamentous** and **spherical particles** are probably unassembled components of Dane particles (Figure 24.-2).

Hepatitis Non-A, Non-B (NANB)

Screening of donated blood for HB_sAg is so effective that most hepatitis from this source is now **non-A, non-B (NANB)**. It is probable that there are two viruses, one transmitted by blood transfusions and another by blood coagulation factors supplied to hemophiliacs. Both diseases resemble hepatitis B. A form of NANB spread by the fecal-oral route resembles hepatitis A and occurs mainly in India and Africa. There is no reliable test at present for NANB in blood.

Table 24-1 Characteristics of Viral Hepatitis

Characteristic	Hepatitis A	Hepatitis B	Delta Hepatitis	Non-A, Non-B (fecal-oral transmission).	Non-A, Non-B (parenteral transmission; may be two viruses involved)
Transmission	Fecal-oral (ingestion of contaminated food and water)	Parenteral (injection of contaminated blood or other body fluids)	Parenteral (host must be coinfected with hepatitis B)	Fecal-oral	Parenteral
Agent	Hepatitis A virus (HAV); single-stranded RNA, no envelope	Hepatitis B virus (HBV); double-stranded DNA, envelope	Delta agent (HDV); single-stranded RNA, envelope	Presumed virus of unknown characteristics	Presumed viruses of unknown characteristics
Incubation period	2 to 6 weeks	4 to 26 weeks	Uncertain	2 to 6 weeks	6 to 25 weeks
Manifestations or symptoms	Mostly sub-clinical; severe cases—fever, headache, malaise, jaundice	Frequently sub-clinical; similar to HAV, but fever, headache absent, and more likely to progress to severe liver damage	Severe liver damage, high mortality rate	Similar to HAV, but pregnant women may have high mortality rate.	Similar to HBV
Chronic liver disease	No	Yes	Yes	No	Yes
Vaccines	Under development	Two types: Heptavax B®, Recombivax HB®	None	None	None

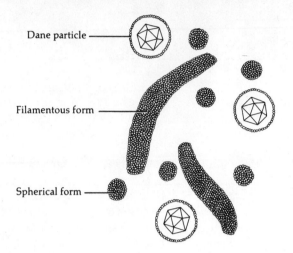

Dane particle

Filamentous form

Spherical form

Figure 24-2 Diagrammatic representation showing the three distinct types of hepatitis B particles discussed in the text.

Delta Hepatitis

The delta agent is a defective virus, unable to cause infection on its own. It reproduces only as a coinfection with the HBV, and causes an acutely worse disease. The delta agent (HDV) has an envelope identical to HBV surrounding a protein core. It is genetically similar to certain plant viruses called viroids.

Viral Gastroenteritis

About 90 percent of viral gastroenteritis is caused by **rotavirus** or **Norwalk agent**. Rotavirus (rota means wheel) is the most common, mostly causing disease in small children. It causes low-grade fever, diarrhea, and vomiting. Major outbreaks of gastroenteritis caused by the Norwalk agent have occurred. Infected persons suffer from gastroenteritis for one to three days.

Fungal Diseases of the Digestive System

Ergotism is caused by the toxin ergot, which is produced by the fungus *Claviceps purpurea* growing on grain. **Aflatoxin** is produced by the mold *Aspergillus flavus* growing on grains, etc. It is toxic to livestock and is suspected to be a human carcinogen.

Protozoan Diseases of the Digestive System

Giardiasis

A flagellated protozoan, *Giardia lamblia*, causes the prolonged diarrheal disease in humans called **giardiasis**. The disease also is characterized by malaise, nausea, weakness, weight loss, and abdominal cramps. Treatment is by metronidazole and quinacrine hydrochloride. The general population has many carriers who may excrete the cyst stage, which is relatively resistant to the chlorine used to disinfect drinking water.

Balantidiasis (Balantidial Dysentery)

A ciliated protozoan, *Balantidium coli*, causes **balantidiasis** (**balantidial dysentery**). Usually mild, it consists of abdominal pain, nausea, vomiting, diarrhea, and weight loss. Humans acquire *B. coli* by ingesting cysts in food or water.

Amoebic Dysentery (Amoebiasis)

The cysts of *Entamoeba histolytica* are ingested with contaminated food and water, and can cause severe **amoebic dysentery**. Severe infections can cause perforation of the intestinal wall and abscesses. For treatment, metronidazole and diiodohydroxyquin are the drugs of choice.

Cryptosporidiosis

A protozoan, *Cryptosporidium*, causes **cryptosporidiosis**, a severe, prolonged diarrhea. Formerly recognized only as a pathogen of calves, in recent years *Cryptosporidium* has been observed in immunosuppressed humans; especially those with AIDS.

Helminthic Diseases of the Digestive System

Tapeworm Infestation

Cysticerci, forms of the **tapeworm** that become encysted in muscles of the intermediate host, are ingested by humans and develop into adult tapeworms (Figure 24-3). They attach to the intestinal wall of the human host and shed eggs with the host's feces. *Taenia saginata* (the beef tapeworm), *Taenia solium* (the pork tapeworm), and *Diphyllobothrium latum* (the fish tapeworm) are the best known forms in the United States. Tapeworms may be 6 or 7 meters in length. Niclosamide is used in treating tapeworm infestations. The pork tapeworm may produce the larval stage in humans, called **cysticercosis**. The infection arises from ingestion of tapeworm eggs. Symptoms can be serious, especially if they develop in the eyes or in the brain.

Hydatid Disease

Echinococcus granulosus, a tapeworm only a few millimeters in length, may infect humans by way of dog feces. The dog becomes infected by eating infected sheep, etc. The tapeworm egg migrates in the human to various tissues and develops into a large hydatid cyst. These are damaging because of their size (see Figure 24-20 in the text) and, if they rupture, possible anaphylactic shock.

Nematode Infections

Pinworms. The most familiar nematode infection in humans is by *Enterobius vermicularis* (**pinworm**). The worm migrates out of the anus of the human host and lays its eggs, causing local itching. Drugs such as pyrantel pamoate or mebendazole are effective treatment.

Hookworms. Hookworm infestations were once common in the southeastern states, mostly caused by *Necator americanus*. The hookworm attaches to the intestinal wall and feeds on blood and tissue. Blood loss can lead to craving for starch or clay soils (pica), which is a symptom of iron deficiency anemia.

Ascariasis. One of the most widespread helminthic diseases is **ascariasis**, caused by *Ascaris lumbricoides*. The worms can be up to about a foot in length, but cause few symptoms as they live on partially digested food in the intestinal tract. Ingested eggs hatch into small, wormlike larva that pass from the bloodstream into the lungs. In the lungs they migrate into the throat and are swallowed, developing into adults in the intestinal tract. Mebendazole is an effective treatment.

Trichinosis. The small roundworm *Trichinella spiralis* causes **trichinosis**. The disease is acquired by ingestion of undercooked meat, usually pork, that contains the encysted larvae. In the intestine of a human the ingested cyst matures into the adult form, producing larvae that invade tissue—especially muscle of the diaphragm and eye. Thiabendazole is an effective drug.

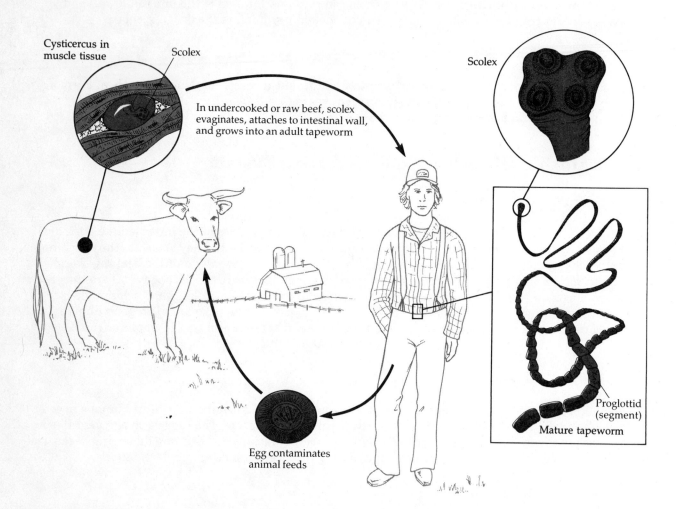

Figure 24-3. Life cycle of the beef tapeworm, *Taenia saginata*. The adult tapeworm lives in the intestine of the human, the definitive host. Tapeworm proglottids and eggs are eliminated with feces and ingested by intermediate hosts such as grazing cattle. The tapeworm eggs hatch, and cysticerci form in the animal's muscles, to be consumed later by humans. The pork tapeworm, *Taenia solium*, has a similar life cycle.

Test

In the matching section, there is only one answer to each question; however, the lettered options (a, b, c, etc.) may be used more than once or, perhaps, not at all.

I. Match

1. Produced from sucrose by *Streptococcus mutans*.

2. Pertaining to inflammation and degeneration of gum and bones supporting teeth.

3. Produced by pathogenic strains of *Staphylococcus aureus*.

4. Disease resulting from ingestion of a toxin.

5. Disease resulting from microbial growth in tissues of the host.

6. Loss of appetite.

a. Periodontal disease

b. Coagulase

c. Intoxication

d. Infection

e. Dental plaque

f. Anorexia

II. Match

1. Affects the parotid glands.

2. Causes diarrhea; same organism that causes gas gangrene.

3. Caused by a small roundworm that encysts its larval form in muscle tissue.

4. MMR vaccine used.

5. A virus commonly causing gastroenteritis.

6. Caused by *Salmonella typhi*.

7. Disease caused by growth of *Salmonella* bacteria in the intestinal tract; incubation time of 12 to 26 hours is normal.

8. Orchitis is a possible complication.

a. Balantidiasis

b. Salmonellosis

c. Norwalk agent

d. Trichinosis

e. Typhoid fever

f. *Clostridium perfringens* gastroenteritis

g. Mumps

III. Match

1. Shigellosis.

2. *Streptococcus mutans*.

3. Most common food poisoning in Japan; halophilic bacteria.

4. Prolonged diarrhea caused by a flagellated protozoan.

5. *Entamoeba histolytica*.

6. Pregnancy and other stresses may activate latent viral infections, which the mother may transmit to the fetus.

7. Coagulase-positive organisms.

a. Dental caries

b. Staphylococcal food poisoning

c. Bacillary dysentery

d. *Vibrio parahaemolyticus* gastroenteritis

e. Cytomegalovirus inclusion disease

f. Giardiasis

g. Amoebic dysentery

IV. Match

1. Infectious hepatitis, usually foodborne.

2. A term associated with the serological classification of *Salmonella*.

3. An infectious viral particle that causes disease only from coinfection with the hepatitis B virus.

4. Mycotoxin.

5. Associated with tapeworm infestation.

a. Aflatoxin
b. Hepatitis A
c. Hepatitis B
d. Non-A, Non-B
e. Delta
f. Kauffman-White
g. Bacteriocinogen
h. Hydatid-cysts

V. Fill-in-the-Blanks

1. Cysticercosis occurs in humans from infection by *Taenia* _____.

2. The epidemic form of cholera is caused by serogroup _____ of *Vibrio cholerae*.

3. The term for skin yellowing due to liver infection is _____.

4. The term for the larval form of tapeworm that becomes encysted in the muscles of the intermediate host is _____.

5. *Diphyllobothrium latum* is the name of the _____ tapeworm.

6. The name for the pork tapeworm is *Taenia* _____.

7. Ergotism is caused by the mycotoxin _____.

8. The species of *Shigella* most common in the United States, which causes a relatively mild dysentery, is *Shigella* _____.

9. Cholera is caused by the organism named _____ *cholerae*.

10. _____ is the toxin produced by the mold *Aspergillus flavus* growing on grains, etc.

11. _____ immunoglobulins in saliva tend to prevent microbial attachment.

12. Dental plaque is formed mostly from _____ produced from sucrose.

13. _____ disease is a collective term for a number of conditions characterized by inflammation and degeneration of the gums, supporting bone, etc.

14. Serological type _____ enterotoxin is responsible for most cases of staphylococcal food poisoning.

15. Pathogenic staphylococci usually ferment _____.

16. In the alimentary canal, the "throat" is the _____, and the "food tube" is the _____.

17. The stomach and the _____ intestine have few microorganisms, but the _____ intestine contains as many as 100 billion bacteria per gram of feces.

18. The Dane particle probably represents an intact _____ virion.

19. *Trichinella* spiralis is a small roundworm that causes _____.

20. For treatment of amoebiasis, one drug of choice is _____; it also is used in treatment of giardiasis.

Key

Matching

I. 1.e 2.a 3.b 4.c 5.d 6.f
II. 1.g 2.f 3.d 4.g 5.c 6.e 7.b 8.g
III. 1.c 2.a 3.d 4.f 5.g 6.e 7.b
IV. 1.b 2.f 3.e 4.a 5.h

Fill-in-the-Blanks

1. *solium*   2. 0:1   3. jaundice   4. cysticeri   5. fish   6. *solium*   7. ergot
8. *sonnei*   9. *Vibrio*   10. aflatoxin   11. IgA   12. dextran   13. periodontal
14. A   15. mannitol   16. pharynx; esophagus   17. small; large   18. Hepatitis B
19. trichinosis   20. metronidazole

25 *Microbial Diseases of the Urinary and Reproductive Systems*

The **urinary system** consists of organs that regulate the chemical composition of the blood and excrete waste products. The **reproductive system** consists of organs that produce gametes for reproduction of the species or to support the developing embryo.

Structure and Function of the Urinary System

The urinary system consists of two kidneys, two ureters, a urinary bladder, and the **urethra** (see Figure 25-1). The kidneys contain **nephrons** that control the solute and waste concentrations in the blood. The **urine** contains urea, uric acid, creatinine, and various salts, together with water. The urine passes down the **ureters** from the **kidneys**, where wastes have been removed from the blood, to the **bladder**. Elimination eventually occurs through the urethra. The urethra in the male also conveys seminal fluid. Valves in ureters prevent backflow to the kidneys from the bladder, an aid in preventing infection. Also, the acidity of urine and the flushing action of excretion aid in preventing infection.

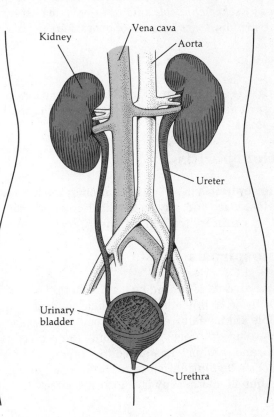

Figure 25-1. Organs of the human urinary system.

Structure and Function of the Reproductive System

The female reproductive system consists of two ovaries, two uterine (fallopian) tubes, the uterus, the vagina, and the external genitals. The ovum (egg) is released from the ovary (a process called ovulation) and passes down the uterine tubes. If it is fertilized in the uterine tube, it becomes implanted in the wall of the uterus. There, it develops into an embryo and fetus. The vagina is a copulatory canal and part of the birth canal. The external genitals are the clitoris, labia, and glands for lubricating secretions. The male reproductive system consists of two testes, which make sperm and male sex hormones; ducts to transport seminal fluid and sperm; and the penis.

Normal Flora of the Urinary and Reproductive Systems

The urine in the bladder and the organs of the upper urinary tract are normally sterile. The urethra contains a resident flora. In infancy, the female vagina is populated by acid-forming lactobacilli, but the acidity becomes more neutral in childhood, and the flora then contains a variety of corynebacteria, cocci, and bacilli. At puberty, acid-forming lactobacilli again become the dominant flora. At menopause, the flora return to that of childhood.

Bacterial Diseases of the Urinary System

Infections of the urinary system are mostly opportunistic. Inflammation of the urethra is called urethritis; of the bladder, cystitis; and of the ureters, ureteritis. When more than 100,000 bacteria per milliliter are found in urine, it usually indicates an infection; normal populations are fewer than 10,000. In females it is not unusual to have the urinary tract contaminated with intestinal bacteria such as *Escherichia coli* and other enteric bacteria. About 35 percent of nosocomial infections occur in the urinary tract because of operations in this area or catherization for draining urine.

Cystitis

Cystitis is an inflammation of the urinary bladder and is more common in females. Gram-negative rods such as *E. coli*, *Proteus vulgaris*, and *Pseudomonas aeruginosa* are frequently involved with both sexes.

Pyelonephritis

Pyelonephritis is an inflammation of the kidneys. Scar tissue can be formed that impairs kidney function. It is generally a complication of infections elsewhere in the body; in females, these usually are lower urinary tract infections. *E. coli* is responsible for about 75 percent of the cases.

Leptospirosis

Leptospirosis is caused by a spirochete, *Leptospira interrogans*. The disease is characterized by chills, fever, and headache. Localization of the pathogens in the kidneys may result in jaundice and possible kidney failure. Leptospirosis is a disease of animals that excrete the organism with the urine—domestic dogs are commonly immunized against it. The organism can pass through mucous membranes and is transmitted to humans and animals by contact with urine-contaminated water. Serological agglutination tests can be used in diagnosis. Penicillin seems to be the most satisfactory drug, but chemotherapy is not always effective.

Glomerulonephritis

Glomerulonephritis, or **Bright's disease**, is an inflammation of the glomeruli (blood capillaries in the kidneys that assist in blood filtration). Glomerulonephritis is an immune-complex disease caused by soluble streptococcal antigens from infections by beta-hemolytic streptococci. The complexes cause inflammation of the glomeruli and kidney damage. The disease is characterized by fever, high blood pressure, and protein and red blood cells in the urine. Antibiotics are effective in combating the original streptococcal infection.

Diseases of the Reproductive System

Bacterial Diseases of the Reproductive System

Diseases of the reproductive system transmitted by sexual activity are called **sexually transmitted diseases (STDs)** or **venereal disease (VD)**.

Gonorrhea

By far the most common reportable communicable disease in the United States is **gonorrhea**, which is caused by a gram-negative diplococcus, **Neisseria gonorrhoeae**. About 1 million cases a year are reported, with about 60 percent occurring in the 15 to 24 age group. Because, to be infective, the bacteria must attach to the epithelial wall of the urethra by means of pili, an experimental vaccine is aimed at these pili. The organism may invade spaces separating mucosal cells found in the oral-pharyngeal area and genitalia, as well as the eyes, and rectum.

Males become aware of gonorrheal infection by painful urination and discharge of pus from the urethra. About 80 percent show these symptoms after a few days incubation. Complications such as sterility resulting from partial blockage of reproductive ducts by scarring may occur.

Females usually are unaware of the infection, but pain can accompany the fairly common occurrence of a spread of the infection to the uterus and uterine tubes (**salpingitis**). Passage of ova to the uterus can be blocked by scar tissue. This can result in sterility or an **ectopic (tubal) pregnancy**. **Pelvic inflammatory disease (PID)** is a term for any extensive bacterial infection of the pelvic region, especially that involving the uterus, uterine tubes, and ovaries.

Complications of gonorrhea may involve the joints, heart (gonorrheal endocarditis), meninges (gonorrheal meningitis), eyes, and pharynx. Gonorrheal arthritis is caused by the growth of the gonococcus in fluids in the joints. Ophthalmia neonatorum is a condition resulting from infection of the eyes of an infant as it passes through the birth canal. Laws usually require administration of silver nitrate or antibiotics at birth to prevent this. Gonorrheal infections also can be transferred by hand contact in adults. Gonorrhea can affect any area of sexual contact. Pharyngeal gonorrhea often resembles the symptoms of a sore throat, and anal gonorrhea can be painful with the discharge of pus.

There is no immunity to reinfection. Penicillin generally is effective for treatment, but resistance is an increasing problem. Tetracycline often is administered because 20 to 25 percent of gonorrhea patients also have concurrent chlamydial infections. Diagnosis is made by identification of the organism in the infectious discharges of both men and women. The organisms are observed as gram-negative cocci in pairs contained within phagocytic leukocytes. The organism survives poorly outside the body and requires special transporting media to keep it alive even for short intervals before cultivating.

Nongonococcal Urethritis (NGU)

Inflammations of the urethra not caused by *Neisseria gonorrhoeae* are called **nongonococcal urethritis (NGU)**, or **nonspecific urethritis**. NGU may be the most common sexually transmitted disease in the United States, but it is not reportable. Many of these infections are caused by chlamydias, obligate intracellular parasites not easily cultured. New tests are available to detect chlamydial infections. Tetracycline is an effective treatment. Other bacteria, such as the mycoplasmas *Ureaplasma urealyticum* or *Mycoplasma hominis*, may cause NGU.

Syphilis

The number of **syphilis** cases in the United States has remained fairly stable in recent years. The causative organism is a gram-negative spirochete, *Treponema pallidum*. Syphilis is transmitted mainly by sexual contact. The incubation period averages about three weeks.

There are several identifiable stages of the disease. The initial symptom of the **primary stage** is a **chancre**, or sore, which usually appears at the site of infection. The fluid formed in the chancre is highly infectious and spirochetes can be seen with darkfield microscopy. In a few weeks this lesion disappears; many women are entirely unaware of it. Serological tests are positive in about 80 percent of patients in the primary stage. Bacteria enter the bloodstream and the lymphatic system and become distributed in the body.

Several weeks after the primary stage, the **secondary stage**, characterized by rashes, appears. The rash occurs on the skin and mucous membranes, and lesions are very infectious. Nonsexual transmission by kissing, dental work, etc., is most likely at this stage. Serological tests become almost uniformly positive. In a few weeks the symptoms of secondary syphilis subside and the **latent period** is entered. The majority of cases do not progress beyond the secondary stage, even without treatment, perhaps because of developing immunity.

After two to four years, the disease is not normally infectious but can be transmitted from mother to fetus. Such **congenital syphilis** can cause damage to the infant's mental development, as well as other neurological disturbances. In fewer than half of the cases, even without treatment, the disease reappears in a tertiary stage; when it does, it usually is after an interval of at least 10 years. Many of the lesions of the **tertiary stage**, called **gummas**, can cause extensive tissue damage. The symptoms may be due to hyperimmune reactions to surviving spirochetes. This stage of syphilis is no longer common.

The **VDRL (Venereal Disease Research Laboratory)** slide test and other rapid screening tests for diagnosis, such as the **rapid plasma reagin (RPR)** card test, detect antibodies, not against the spirochete itself, but reagin-type antibodies. The production of these is stimulated by lipid material formed as a response to infection by the spirochete. The antigen used in these slide tests is an extract of beef heart, which contains lipids similar to those formed by the body that originally stimulated the reagin production. Confirmation of a positive screening test can be made by **fluorescent treponemal antibody absorption (FTA-ABS)** tests. In these tests an avirulent cultivated strain of the spirochete is allowed to react with antibodies present in the patient's blood. Benzathine penicillin, a long-lasting formulation, has continued to be effective in the treatment of syphilis, especially in the primary stage.

Gardnerella Vaginalis

Vaginitis, infections of the vagina, caused by the fungus *Candida albicans*, the protozoan *Trichomonas vaginalis*, or the bacterium *Gardnerella vaginalis*. *C. albicans* usually is an opportunist, while *T. vaginalis* and *G. vaginalis* usually are sexually transmitted. *G. vaginalis* infection involves interaction with anaerobic vaginal bacteria. Diagnosis is based on a fishy odor, a vaginal pH above 5, and "clue cells," sloughed-off vaginal epithelial cells covered with gram-negative rods, in the discharge. Metronidazole, which eliminates the anaerobic bacteria, is used for treatment.

Lymphogranuloma Venereum

Chlamydia trachomatis, the probable major cause of NGU and the cause of trachoma, also is responsible for **lymphogranuloma venereum**. The disease is found in much of the tropical world and in the southeastern United States. Infection causes the regional lymph nodes to become enlarged and tender. Scarring of the lymph drainage ducts occasionally leads to massive enlargement of the male genitalia and rectal narrowing in females. Chlamydias, if properly stained, may be seen as inclusions in pus. Tetracycline antibiotics generally are effective in treatment.

Chancroid (Soft Chancre) and Granuloma Inguinale

Chancroid (**soft chancre**) and **granuloma inguinale** largely are tropical diseases. In chancroid, a painful ulcer forms on the genitalia and spreads to adjacent lymph nodes. The causative organism is *Hemophilus ducreyi*. Erythromycin and trimethoprin-sulfamethoxazole are effective in treatment.

Granuloma inguinale is characterized by an initial chancre on or about the genitals, which may spread and destroy large areas of genital tissue. The disease is most common in India, the Caribbean, and Southeast Asia. The causative organism is *Calymmatobacterium granulomatis*, which can be seen in scrapings of lesions, along with mononuclear cells called Donovan bodies. Gentamicin and chloramphenicol are the drugs of choice.

Viral Diseases of the Genital System

Genital Herpes

Herpes simplex virus type 2, and sometimes type 1, causes a sexually transmitted disease called **genital herpes**. The lesions cause a burning sensation, and vesicles appear. The lesions are infectious; there is a serious danger of infection to an infant at birth, **neonatal herpes**. Such an infection can be fatal or cause serious neurological damage.

The virus may enter a latent stage in nerve cells and reappear at intervals, much like cold sores. There is no cure at this time for genital herpes, although oral and topical administration of acylovir alleviates symptoms and decreases the frequency of recurrence in many cases.

Genital Warts

Warts are caused by viruses, and sexual transmission of genital warts is common. The papillomaviruses that cause warts also are factors in cervical cancer.

AIDS

AIDS is caused by a virus that is often sexually transmitted; it is discussed in Chapter 18 because it affects the immune system.

Fungal Disease of the Reproductive System

Candidiasis

Candida albicans is a yeastlike fungus that grows on mucous membranes. Infection of the mucous membrane of the vagina is **vulvovaginal candidiasis**. There is irritation, severe itching, a thick, yellow, cheesy discharge, and a yeasty odor. Predisposing conditions include pregnancy, diabetes, certain tumors, and treatment with immunosuppressive drugs or broad spectrum antibiotics. Diagnosis is made by microscopic identification of the fungus, and treatment consists of topical application of clotrimazole and miconazole.

Protozoan Disease of the Reproductive System

Trichomoniasis

The protozoan *Trichomonas vaginalis* is a fairly normal inhabitant of the female vagina and the male urethra. Trichomoniasis results when the acidity of the vagina is disturbed and the protozoans outgrow the normal microbial flora. Males rarely have any symptoms; females exhibit a purulent discharge with a disagreeable odor accompanied by irritation and itching. Diagnosis is made by microscopic examination of the discharge. Metronidazole is used in treatment.

Test

In the matching section, there is only one answer to each question; however, the lettered options (a, b, c, etc.) may be used more than once or, perhaps, not at all.

I. Match

1. Most common reportable communicable disease in the U.S.

2. The VDRL test useful in diagnosis.

3. Caused only by a protozoan.

4. Most of the cases probably are caused by chlamydias.

5. In diagnosis, the causative organism of this disease is seen as pairs of gram-negative cocci contained within phagocytic leukocytes.

6. *Treponema pallidum.*

7. Caused by a spirochete.

8. Metronidazole has been an effective treatment.

9. Caused by a virus.

a. Nongonococcal urethritis

b. Gonorrhea

c. Genital herpes

d. Syphilis

e. Trichomoniasis

II. Match (stages of syphilis)

1. Characteristic symptoms mainly are rashes; serological tests at this stage almost uniformly positive.

2. The disease is not normally infectious during this stage, but it can be transmitted from mother to fetus.

3. The usual symptom characteristic of this stage is the chancre.

4. Lesions in this stage can cause extensive damage; symptoms may be due to hyperimmune reactions.

a. Primary

b. Latent

c. Tertiary

d. Secondary

III. Match

1. Site of ectopic pregnancies.

2. Where the ovum is implanted to develop into a fetus.

3. Channel for eventual elimination of urine.

4. Carries urine from the kidneys to the urinary bladder.

5. Site of cystitis.

a. Ureter

b. Bladder

c. Urethra

d. Uterus

e. Uterine tube

f. Urine

IV. Match

1. Caused by *Chlamydia trachomatis*.

2. Bright's disease.

3. Immune-complex disease caused by soluble streptococcal antigens.

4. Caused by a spirochete.

a. Glomerulonephritis

b. Leptospirosis

c. Granuloma inguinale

d. Lymphogranuloma venereum

V. Fill-in-the-Blanks

1. The ovum is released from the ovary in a process called _____.

2. Inflammation of the urethra is called _____.

3. Inflammation of the ureters is called _____.

4. The spread of an infection such as gonorrhea to the uterus and uterine tubes is called _____.

5. Infection of the eyes of the newborn by the gonococcus at the time of birth is _____.

6. To prevent infection of the eyes of the newborn with the gonococcus from infected mothers, the law requires administration of _____ or antibiotics.

7. Slide tests to screen for syphilis detect _____ - type antibodies.

8. *Hemophilus ducreyi* is the cause of _____.

9. Genital herpes type _____ is the cause of genital herpes.

10. The uterine tubes also are known as the _____ tubes.

11. When more than _____ bacteria per milliliter are found in urine, it usually indicates an infection.

12. Pyelonephritis is an inflammation of the _____.

13. Domestic dogs are commonly immunized against a disease usually spread by contact with urine-contaminated water.

14. Gonorrheal _____ is caused by growth of the gonococcus in fluids in the joints.

15. To be infective, the organism causing gonorrhea attaches to the epithelial walls of the urethra by _____.

16. Syphilis transmitted by the mother to the fetus is known as _____ syphilis.

17. Infection of the mucous membrane of the vagina by *Candida albicans* is _____.

18. Rubbery lesions of the tertiary stage of syphilis, called _____, sometimes cause extensive tissue damage.

19. In females, the urinary tract can become contaminated with intestinal bacteria such as _____.

20. Another name for chancroid is _____.

21. A drug useful in alleviating symptoms of genital herpes is _____.

Key

Matching

I.   1.b 2.d 3.e 4.a 5.b 6.d 7.d 8.e 9.c
II. 1.d 2.b 3.a 4.c
III. 1.e 2.d 3.c 4.a 5.b
IV. 1.d 2.a 3.a 4.b

Fill-in-the-Blanks

1. ovulation   2. urethritis   3. ureteritis   4. salpingitis   5. opthalmia neonatorum
6. silver nitrate   7. reagin   8. chancroid   9. 2   10. fallopian   11. 100,000
12. kidneys   13. leptospirosis   14. arthritis   15. pili   16. congenital
17. vulvovaginitis candidiasis (vaginitis)   18. gummas   19. *Escherichia coli*
20. soft chancre   21. acyclovir

26 *Soil and Water Microbiology*

Relatively few microorganisms are harmful; in fact, bacteria and other microorganisms are intimately involved in the maintenance of life on earth. Microbes are found in widely varied habitats on earth, from the thin atmosphere at high altitudes to the deepest oceans; from the frozen ices to the boiling hot springs.

Some microbes are **parasites** that prey on others. Some form a **symbiosis**, such as the association between fungi and algae in lichen. A common relationship is **commensalism**, where one organism benefits without affecting the other. One example of commensalism is partial degradation of cellulose by some so that the resulting glucose can be used by others. In **cometabolism** the degrading organisms are not benefitted.

Soil Microbiology and Cycles of the Elements

The Components of Soil

The uppermost layer of soil, **topsoil**, is the most important for living organisms.

Minerals. Most soil consists of a mixture of rock and mineral fragments formed by the weathering of preexistent rock.

Organic matter. Organic matter is derived from the remains of various living organisms and their waste products. Much of organic matter is of plant origin; much comes from the vast numbers of microorganisms and small animals in the soil. The anaerobic environment of swamps and bogs leads to slower decomposition and therefore a higher organic matter content—about 95 percent compared to 2 to 10 percent found in agricultural soils. A considerable part of soil organic matter is **humus**, dark residual organic material relatively resistant to decay. Organic matter gives soil its spongy nature, prevents formation of heavy crusts, and increases the proportion of the pore spaces in soil, which then increases aeration and water retention.

Water and Gases. Water is found between the soil particles and adheres to them. Various inorganic and organic constituents of soil are dissolved in the soil water and made available to living inhabitants of the soil. Soils, because of biological respiratory activity, contain a higher concentration of carbon dioxide and a lower proportion of oxygen than the atmosphere.

Organisms. Microorganisms in the soil exist in a condition of near starvation with low reproductive rates. When usable nutrients are added to soil, the microbial populations and their activity rapidly increase, but return to normal levels when the nutrients are depleted. The most numerous organisms in soil are **bacteria**. A typical garden soil will have millions per gram, mainly in the top few centimeters. Numbers determined by the usual plate count methods probably greatly underestimate the actual amount. **Actinomycetes**, although bacteria, are usually considered separately when enumerating soil bacteria. Found in soil in large numbers, they produce **geosmin**, the gas that gives soil its characteristic musty odor. The actual **biomass** (weight) of actinomycetes is probably about that of conventional bacteria. Plate counts may enumerate mainly the asexual spores and mycelial fragments of these filamentous organisms. The genus *Streptomyces* produces many of our antibiotics.

Fungi are found in soil in fewer numbers than bacteria and actinomycetes. Estimates of the biomass of fungi, however, indicate that it probably equals that of bacteria and actinomycetes combined. Molds greatly outnumber yeasts in soil. **Algae** and **cyanobacteria** sometimes form abundant growth on the surface of moist soils, but also are found in desert soils. They are located mainly in the surface layers. **Protozoan** populations in the soil tend to rise and fall with their food supply, which is mainly bacteria.

Pathogens in soil. Human pathogens find the soil an alien, hostile environment. Even relatively resistant enteric pathogens such as *Salmonella* species survive only a few weeks or months. However, endospore-forming bacteria such as *Bacillus anthracis*, the cause of anthrax in animals, survive in soil and are ingested by grazing animals. *Clostridium tetani* (tetanus), *Clostridium botulinum* (botulism), and *Clostridium perfringens* (gas gangrene) are endospore-forming soil organisms that are pathogenic when introduced into anaerobic food or wounds, where they produce toxins.

Plant pathogens, the cause of rusts, smuts, blights, and wilts that affect plants, are often caused by fungi that pass part of their life cycle in soil. **Insect pathogens** such as *Bacillus thuringiensis* (whose spores and toxins are sold commercially to control some insect larvae) also are found in soil.

Microorganisms and Biogeochemical Cycles

Perhaps the most important role of soil microorganisms is their participation in biogeochemical cycles; that is, the recycling of certain chemical elements so that they can be used over again.

Carbon Cycle

The first step in the **carbon cycle** is the utilization of atmospheric carbon dioxide (about 0.03 percent of the atmosphere) in photosynthesis to produce organic matter. Eventually this organic matter, from green plants or creatures living directly or indirectly off plants, is changed by decay into carbon dioxide and released again to the atmosphere. Much carbon dioxide also is released by the burning of fossil fuels. There is considerable concern that accumulations of carbon dioxide in the atmosphere will cause world temperatures to increase—the **greenhouse effect**.

Nitrogen Cycle

Nitrogen is needed by all organisms for the synthesis of protein, nucleic acids, and other nitrogen-containing compounds (Figure 26-1). Molecular nitrogen (N_2) comprises about 80 percent of the earth's atmosphere. As nitrogen-containing organic matter enters the soil (much of the nitrogen is in protein form), it undergoes decay by microorganisms. Decomposition of proteins releases amino acids. Ammonia (NH_3) is formed by the removal of amino groups from amino acids—called **ammonification**. In **nitrification**, ammonia is oxidized to yield energy by autotrophic bacteria (see Figure 26-1). During nitrification, *Nitrosomonas* bacteria convert ammonia to nitrites (NO_2), and *Nitrobacter* bacteria oxidize the nitrites to nitrates (NO_3), the form preferred by plants as a nitrogen source. **Denitrification** is an anaerobic process by which nitrates are used as an electron acceptor in place of oxygen (an example of anaerobic respiration). *Pseudomonas* species appear to be the most important of several groups involved. The result is, by a series of steps, formation of nitrogen gas, which returns to the atmosphere. This process can be an important means of nitrogen loss in soil.

    Nitrogen fixation is the ability to use nitrogen directly from the atmosphere, an accomplishment of only a few bacteria or cyanobacteria. There are several types of **nonsymbiotic** (free-living) bacteria capable of nitrogen fixation. Among them are the aerobic *Azotobacter* and *Beijerinckia*. Some anaerobic species of *Clostridium* such as *Clostridium pasteurianum* and a number of species of the photosynthetic cyanobacteria fix nitrogen. The cyanobacteria are important in rice paddies of the Orient, where they form a nitrogen-fixing symbiosis with *Azolla* water plants. The **nitrogenase** (nitrogen-fixing) enzyme system operates only anaerobically and must be protected from oxygen; cyanobacteria carry it in **heterocysts**. Other nonsymbiotic nitrogen-fixing bacteria include certain species of the *Klebsiella*, *Enterobacter* and *Bacillus* genera, as well as the photoautotrophic *Rhodospirillum* and *Chlorobium*. In soil, most of the heterotrophic nonsymbiotic nitrogen-fixing bacteria usually lack sufficient carbohydrates needed for energy to reduce nitrogen to ammonia. They seldom fix the amounts that can be observed under laboratory conditions. Nevertheless, they can make important contributions to the nitrogen economy of areas such as grasslands, forests, and the arctic tundra.

Symbiotic nitrogen-fixing bacteria are agriculturally more important. The genera *Rhizobium* and *Bradyrhizobium* infect the roots of leguminous plants such as soybeans, beans, peas, peanuts, alfalfa, and clover. These bacteria are specific for a particular leguminous species. They attach to a root hair and cause formation of an **infection thread** passing into the root itself. *Rhizobium* and *Bradyrhizobium* entering root cells by the thread become enlarged (**bacteroids**). The root cells proliferate to form a **root nodule** packed with bacteroids. Nitrogen is fixed there, with the plant furnishing anaerobic conditions and growth nutrients and the bacteria fixing the nitrogen (Figure 26-2). Similar examples occur in nonleguminous plants such as alder trees, for which an actinomycete is the microorganism. **Lichens**, a symbiosis between a fungus and an alga or cyanobacterium, may be a source of nitrogen when one symbiont is a nitrogen-fixing cyanobacterium.

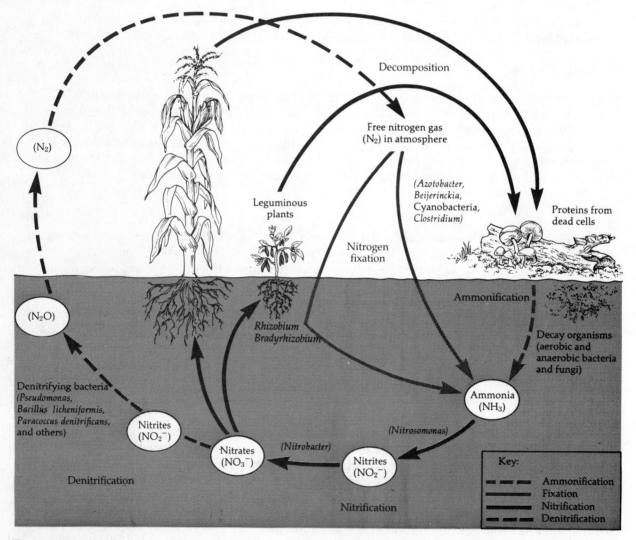

Figure 26-1 The nitrogen cycle.

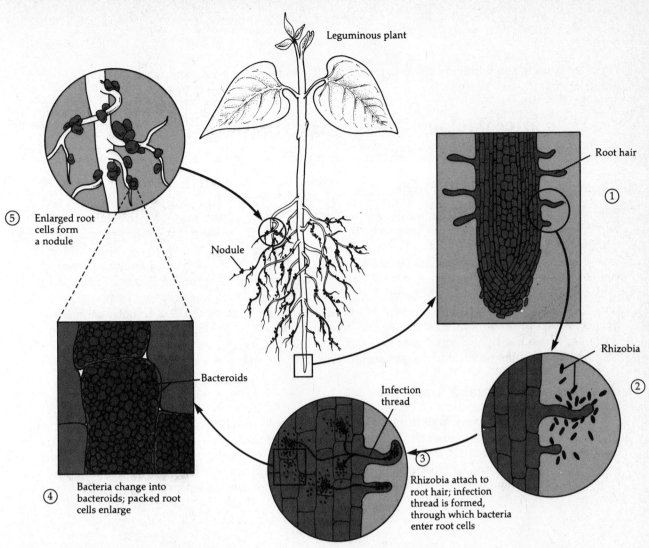

Leguminous plant

⑤ Enlarged root cells form a nodule

Nodule

Root hair

①

Rhizobia

②

Bacteroids

Infection thread

③

④ Bacteria change into bacteroids; packed root cells enlarge

Rhizobia attach to root hair; infection thread is formed, through which bacteria enter root cells

Figure 26-2 Formation of a root nodule.

The Role of Mycorrhizae. There are two types of mycorrhizal fungi: **endomycorrhiza (vesicular-arbuscular mycorrhiza)**, and **ectomycorrhiza**. Both types function like root hairs on plants, extending the surface area through which the plant can absorb nutrients, especially phosphorus. Endomycorrhizae form **vesicles** (oval bodies that are most likely storage structures) and **arbuscules** (bushy structures inside plant root cells that break down and release nutrients to the plants). Ectomycorrhizae infect mainly trees such as pine or oak, forming a mycelial **mantle** over smaller roots. Truffles, a food delicacy, are ectomycorrhizae.

Other Biogeochemical Cycles

Sulfur and several other elements undergo cyclic changes similar to carbon and nitrogen. These cycles, and other transformations mediated by microorganisms, make minerals available to plants in soluble form.

Degradation of Synthetic Chemicals in the Soil

Pesticides and other synthetic chemicals are usually degraded, with varying rapidity, by soil microorganisms. DDT is an example of a **recalcitrant** (resistant to biodegradation) pesticide. Small changes in the molecule can be important. For example, the addition of a single chlorine atom to the ring structure of the herbicide 2,4-D to form 2,4,5-T creates a chemical that is much more recalcitrant.

Aquatic Microbiology and Sewage Treatment

Aquatic microbiology refers to the study of microorganisms in lakes, rivers, estuaries, and oceans.

Freshwater Microbial Flora

High nutrient levels in water generally are reflected by high microbial numbers. Microbial populations of freshwater bodies tend to be differentiated by the availability of oxygen and light (Figure 26-3). The **littoral zone** is the region along the shore that has rooted vegetation and light that penetrates to the bottom. The **limnetic zone** consists of the surface away from the shore. The **profundal zone** is the deeper water under the limnetic zone, and the **benthic zone** is the sediment at the bottom. Light is important because photosynthetic algae are the main source of organic matter (energy) for the lake. These are the primary producers and are located in the limnetic zone. The deeper waters of the profundal zone are low in oxygen and light. There, purple and green sulfur bacteria, which use light filtered past algae at higher levels, are found. In this zone are *Desulfovibrio*, which produce hydrogen sulfide from sulfate, and the methane-producing bacteria.

Seawater Microbial Flora

The open ocean is relatively high in osmotic pressure, low in nutrients, and has a pH above optimum. Bacterial populations there are comparatively low. Much of the microscopic life is photosynthetic diatoms and other algae that make up the **phytoplankton community**, the basis of the oceanic food chain. Krill, shrimplike creatures, feed on phytoplankton and are important food supply for larger sea life, although many fish and whales feed directly on phytoplankton.

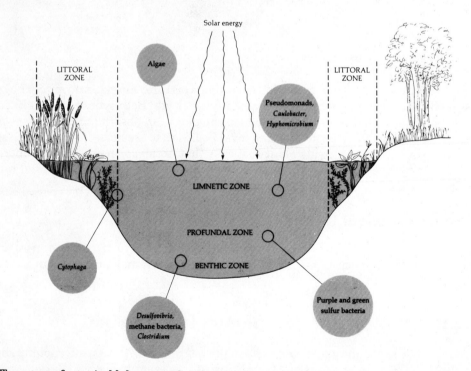

Figure 26-3 The zones of a typical lake or pond and some representative micro-organisms of each zone. The microorganisms fill niches that vary in light, nutrients, and oxygen availability.

The Role of Microorganisms in Water Quality

Water Pollution

Transmission of Infectious Diseases. Many diseases are perpetuated by the **fecal-oral route** of transmission, in which a pathogen shed in human feces contaminates water or food and is later ingested. Examples are typhoid fever, cholera, hepatitis A, and infections by *Giardia lamblia*. Many helminthic diseases such as schistosomiasis are spread among persons coming into contact with contaminated water, not by ingestion; a larval form penetrates the skin.

Chemical Pollution

Many industrial and agricultural chemicals enter water and are resistant to degradation. Some become concentrated in biological organisms, as happens with fat-soluble pesticides and metallic mercury. Mercury is converted by benthic bacteria to a soluble form that is taken up by fish. Eating such fish can cause neurological damage.

 Eutrophication, a term meaning "well nourished," can occur when nitrogen and phosphorus, in particular, enter waters. Algae get carbon from atmospheric carbon dioxide and energy from light. Therefore, algae and non-nitrogen-fixing cyanobacteria require only relatively small amounts of nitrogen and phosphorus for heavy growth (blooms). Many cyanobacteria fix their own nitrogen; for them phosphorus is the only ingredient lacking. The death of these organisms results in their degradation by bacteria, a process that depletes the oxygen in water and hastens the filling of the lake.

 Coal-mining wastes often are high in sulfur content, mostly iron sulfides (FeS_2). Bacteria such as *Thiobacillus ferrooxidans* obtain energy from the oxidation of the ferrous ion (Fe^{2+}) and convert the sulfides into sulfates. These enter the streams as sulfuric acid, damaging aquatic life and causing formation of insoluble yellow iron hydroxides. These organisms also are used in mining to leach minerals such as uranium from ores in a soluble form.

Tests for Water Purity

Microbiological tests for water safety are aimed at detection of **indicator organisms**. In the United States these usually are **coliform bacteria** (aerobic or facultative anaerobic, gram-negative, non-endospore-forming rod-shaped bacteria that ferment lactose with acid and gas formation within 48 hours at 35°C). Coliforms are normal inhabitants of human intestines, and their presence is considered an indication of human wastes entering the water. *Escherichia coli* is the predominant **fecal coliform**. Some coliforms, common soil and plant inhabitants, are **nonfecal coliforms** and are less important as indicators.

 There are two basic methods for the detection and enumeration of coliforms. These are summarized in Figure 26-8 in the text. The first is the **multiple tube fermentation technique**. Fifteen tubes of lactose broth are inoculated by a specified dilution series of water samples. A positive presumptive test such as this can be confirmed (confirmed test) by streaking samples from the test onto differential media such as eosin methylene blue (EMB) agar (which contains lactose). Distinctive colonies with dark centers or a metallic sheen are formed by coliforms.

 Coliforms also may be detected by the **membrane filter method**. Bacteria are retained on the surface of the membrane filter and will form colonies of a distinctive appearance when the membrane is transferred to a pad soaked with a suitable nutrient medium.

Water Treatment

If turbid, water can be held in a reservoir to allow some particulate matter to settle out. The water then undergoes a **flocculation treatment**, by addition of aluminum potassium sulfate (alum), for example. This chemical forms a floc, which carries suspended colloidal material with it as it settles. The next step is passage through beds of sand (**sand filtration**), which removes about 99 percent of the bacteria and viruses remaining at this stage. Microorganisms are trapped mostly by surface adsorption onto the sand grains. Activated charcoal may someday replace or supplement sand filtration; it has the added advantage of removing chemical pollutants as well. Water is then **chlorinated** to kill microorganisms before distribution. **Ozone** is considereed a possible replacement for chlorination. Ozone (O_3) is generated by electricity at the site of treatment. Arrays of **ultraviolet** tube lamps also can be used to disinfect water.

Sewage Treatment

After water is used, it becomes sewage, a term that includes household washing wastes, industrial wastes, and rainwater from street drains. Sewage is mostly water with as little as 0.03 percent particulate matter.

Primary Treatment

The usual first step in sewage treatment is **primary treatment**. Settling chambers remove sand, etc.; skimmers remove floating oil and grease; floating debris is ground. Following this, solid matter is allowed to settle out in sedimentation tanks. Sewage solids collecting on the bottom (40 to 60 percent of suspended solids are removed here) are called **sludge**. Sludge is removed periodically and the effluent (liquid flowing out) may undergo secondary treatment.

Biochemical Oxygen Demand

Biochemical oxygen demand (BOD) is a measure of the biologically degradable organic matter in water, as determined by measuring the amount of oxygen required by bacteria to metabolize it. A classic method is to follow for five days the lowering of oxygen levels in water samples in a sealed bottle. The more oxygen used up as the bacteria utilize the organic matter in the sample, the greater the BOD.

Secondary Treatment

Primary treatment removes 25 to 35 percent of the BOD; the remainder is largely in the form of dissolved organic matter. **Secondary treatment** involves high aeration to encourage aerobic microorganisms to oxidize dissolved organic matter to carbon dioxide and water. In the **activated sludge system**, air is added to aeration tanks. This encourages the growth of organisms that rapidly oxidize organic matter, and causes formation of a suspended flocculant sludge. After several hours, the sludge is passed to a tank where it settles to the bottom. About 20 percent of the sludge is recycled to the activated sludge tanks as a "starter" for the next sewage batch (this is the basis for the term activated sludge). If the sludge floats, the phenomenon is called **bulking,** and is a common problem. The growth of filamentous bacteria such as *Sphaerotilus natans* is a possible cause of bulking. Activated sludge systems remove 75 to 90 percent of the BOD. **Trickling filters** is a secondary treatment method whereby sewage is sprayed over a bed of rocks 2 to 4 inches in diameter and about 6 feet deep. A slime layer composed of aerobic microorganisms that aerobically decompose dissolved organic matter in the passing sewage soon forms on the rocks. Trickling filters remove only 80 to 85 percent of the BOD, but they are less troublesome to operate and less sensitive to toxic sewage and overloads.

Sludge Digestion

Sludge accumulating from primary and secondary treatment can be treated in **anaerobic sludge digesters**. The first stage of activity in this process is a fermentation forming organic acids and carbon dioxide. Microorganisms then produce hydrogen and carbon dioxide from these organic acids.

The hydrogen and carbon dioxide are used by anaerobic methane-producing bacteria to form methane. Most methane is derived from the energy-yielding reduction of carbon dioxide by hydrogen gas:

$$CO_2 + 4H_2 \longrightarrow CH_4 + 2H_2O$$

Other methane-producers split acetate to yield carbon dioxide and methane:

$$CH_3COOH \longrightarrow CH_4 + CO_2$$

Much of the sewage is, therefore, converted to methane and carbon dioxide. Residual solids are pumped to shallow drying beds or to filters to reduce the volume. This sludge is disposed of in several ways, including use as a soil conditioner.

Septic Tanks

Septic tanks settle out suspended solids, which are periodically pumped out, in a holding tank. The effluent flows through a system of perforated piping into a soil drainage field, where it enters the soil and is decomposed by soil microorganisms.

Oxidation Ponds

Many small communities and many industries use **oxidation ponds** (also called **lagoons** or **stabilization ponds**). Many incorporate a two-stage system. The first stage is deep and anaerobic, and settles out sludge. Effluent is then pumped into other ponds that are shallow enough to be aerated by wave action. The growth of algae is encouraged in order to produce oxygen to help maintain aerobic conditions in such high organic matter loads. Bacterial action in decomposing organic matter generates the carbon dioxide used by the algae in photosynthesis in the shallow waters.

Tertiary Treatment

Primary and secondary treatments do not remove all of the biodegradable organic matter. Secondary treatment effluent contains about 50 percent of the original nitrogen and 70 percent of the phosphorus. These proportions can have a great impact on a lake's ecosystem (see earlier discussion of eutrophication). **Tertiary treatment** removes almost all of the BOD, nitrogen, and phosphorus. Nitrogen is converted to ammonia and evaporated into the air or converted to nitrogen gas by denitrifying bacteria. Tertiary treatment is very expensive and is used at this time only in highly developed lake areas where eutrophication must be avoided even at high costs.

Test

In the matching section, there is only one answer to each question; however, the lettered options (a, b, c, etc.) may be used more than once or, perhaps, not at all.

I. Match

1. A gas that gives soil its characteristic odor.

2. Heavy growth of algae.

3. An accumulation of dark organic matter that is relatively resistant to further degradation.

    a. Humus
    b. Topsoil
    c. Geosmin
    d. Blooms
    e. Biomass

II. Match

1. *Rhizobium* species important in this process.

2. *Pseudomonas* species important in this process.

3. *Azotobacter* species important in this process.

4. *Nitrobacter* species important in this process.

5. Forms a mantle over small roots.

6. Forms arbuscules inside plant root cells.

7. Truffles.

    a. Symbiotic nitrogen fixation
    b. Nonsymbiotic nitrogen fixation
    c. Nitrification
    d. Denitrification
    e. Ammonification
    f. Endomycorrhiza
    g. Ectomycorrhiza

III. Match
(aquatic zones)

1. The zone in the region along the shore that has rooted vegetation and light that penetrates to the bottom.

2. The deeper water under the limnetic zone.

3. Bottom sediments.

4. The water surface away from shore.

5. Where the primary producers are mainly located.

6. Where the methane-producing bacteria are found.

    a. Limnetic
    b. Benthic
    c. Littoral
    d. Profundal

IV. Match (sewage treatments):

1. Removes almost all BOD, nitrogen, and phosphorus.

2. Methane-producing bacteria an important element.

3. Mainly designed to encourage growth of aerobic bacteria to oxidize dissolved organic matter to carbon dioxide and water.

4. *Zoogloea* bacteria form floc in one method.

a. Primary

b. Secondary

c. Anaerobic sludge digestion

d. Tertiary

V. Match

1. A morphological form of *Rhizobium*.

2. A water plant that may have a symbiosis with nitrogen-fixing cyanobacteria.

3. A possible substitute for chlorination treatment of water.

4. Shrimplike creature found in the ocean.

5. The filamentous organism *Sphaerotilus natans* may be a factor in this.

a. Krill

b. Bacteroids

c. Lichen

d. *Azolla*

e. Phytoplankton

f. Ozone

g. Bulking

VI. Match

1. Human and animal pathogen found in soil.

2. Insect pathogen found in soil.

3. Fecal coliform.

4. Conversion of ammonia to nitrite.

5. Nonsymbiotic nitrogen fixation involves this bacterium.

a. *Bacillus thuringiensis*

b. *Bacillus anthracis*

c. *Nitrosomonas*

d. *Beijerinckia*

e. *Escherichia coli*

VII. Fill-in-the-Blanks

1. The anaerobic environment of swamps and bogs leads to slower decomposition and therefore a higher organic matter content, about _____ percent.

2. The most numerous organisms in soil are _____.

3. Many of our antibiotics are produced by soil organisms of the genus _____.

4. Estimates of the biomass of _____ in the soil indicate that it probably equals that of bacteria and actinomycetes combined.

5. Carbon dioxide is about _____ percent of the atmospheric gases.

6. Molecular nitrogen is about _____ percent of the atmospheric gases.

7. The form of nitrogen preferred by most plants as a nitrogen source is _____.

8. The removal of amino groups from amino acids to form ammonia is called _____.

9. Denitrification is a process by which nitrates are used as an electron acceptor in place of _____.

10. The *Cyanobacteria* protect their nitrogenase systems from oxygen in _____.

11. *Rhizobium* bacteria infect a plant by attaching to a root hair and causing formation of an _____.

12. _____ are a symbiosis between a fungus and an alga or cyanobacterium.

13. Pesticides resistant to biodegradation are described as being _____.

14. The symbiont of the nitrogen-fixing alder tree is an _____.

15. Many cyanobacteria fix nitrogen, and for them the only requirement for heavy growth in most lake waters is the addition of small amounts of _____.

16. Streaking out the results of a presumptive test for coliforms onto EMB agar is part of the _____ test.

17. Addition of alum to water in purification is called _____ treatment.

18. In sand filtration, the microorganisms are trapped mostly by _____ onto the sand grains.

19. Sewage solids collecting on the bottom are called _____.

20. Biochemical oxygen demand is a measure of the biologically degradable _____ in water as determined by the amount of _____ required by bacteria to metabolize it.

21. When the sludge fails to settle out properly in activated sludge processing, it is called

_____.

22. In oxidation ponds the growth of algae is often encouraged in order to produce

_____.

Key

Matching

I.   1.c 2.d 3.a
II. 1.a 2.d 3.b 4.c 5.g 6.f 7.g
III. 1.c 2.d 3.b 4.a 5.c 6.b
IV. 1.d 2.c 3.b 4.b
V.   1.b 2.d 3.f 4.a 5.g
VI. 1.b 2.a 3.e 4.c 5.d

Fill-in-the-Blanks

1. 95   2. bacteria   3. *Streptomyces*   4. fungi   5. 0.03   6. 80   7. nitrate
8. ammonification   9. oxygen   10. heterocysts   11. infection thread   12. Lichens
13. recalcitrant   14. actinomycete   15. phosphorus   16. confirmed or confirmatory
17. flocculation   18. adsorption   19. sludge   20. organic matter; oxygen   21. bulking
22. oxygen

27 *Applied Microbiology and Biotechnology*

Food Preservation and Spoilage

Modern civilization and the support of large populations could not exist were it not for effective methods of preserving food. Drying food or adding salt or sugar lowers the available moisture and prevents spoilage. The acidity found in the natural fermentation of milk or vegetable juices prevents growth of many spoilage bacteria. Comparatively, preservation of foods by heat (**canning**) and refrigeration requires considerable technical sophistication. Canned foods today undergo what is called **commercial sterilization**, the minimum processing necessary to destroy the endospore-forming pathogen *Clostridium botulinum*. To ensure this result, the **12D treatment** is applied, by which a theoretical population of botulism bacteria would be decreased by 12 logarithmic cycles. (10^{12} endospores would be reduced to only one.)

Spoilage of Canned Food

Thermophilic anaerobic spoilage is possible in low-acid canned foods. The can usually is swollen from gas production, and the contents have a lowered pH and a sour odor. The cause is thermophilic clostridia, which survived commercial sterilization because of unusually resistant endospores. In **flat sour spoilage**, caused by thermophilic organisms such as *Bacillus stearothermophilus*, the can is not swollen by gas production. Both of these types of spoilage occur with storage at higher than normal temperatures. The thermophilic organisms involved will not grow at normal storage temperatures. Canned food spoilage by **mesophilic bacteria** is due to can leakage or underprocessing. The former is more likely to result in spoilage by non-endospore-formers, and the latter spoilage by endospore-formers.

   Contamination by leakage can be caused by sucking cooling water past the sealant in the crimped lid as the can cools. These types of spoilage often result in odors of putrefaction in high-protein foods. Botulism is a possible consideration in mesophilic spoilage. Some acidic foods such as tomatoes and fruits are preserved by heating at temperatures below the boiling point. This type of preservation is possible because organisms such as molds, yeasts, and occasional species of acid-tolerant non-endospore-forming bacteria (the only organisms likely to grow in such foods) are easily killed. One problem with such acidic foods can be the heat-resistant **sclerotia** (specialized resistant bodies) of certain molds, which can survive temperatures of 80°C for a few minutes. Other problems are **heat-resistant ascospores** of the mold *Byssochlamys fulva*, and the endospore-forming *Bacillus coagulans*, a bacterium capable of growth at a pH of almost 4.0.

Aseptic Packaging

The container used in **aseptic packaging** is usually made of laminated paper or plastic that cannot tolerate conventional heat treatment. Rolls of packaging materials are fed into a machine that sterilizes the material in hot hydrogen peroxide. The containers are formed while still in the sterile environment and filled with conventionally heat-sterilized foods.

Low-Temperature Preservation

Low temperatures slow the reproduction time of microorganisms; however, some molds and bacteria can grow at a significant rate even below the freezing point of water. A properly set refrigerator maintains a temperature of 0 to 7°C. Although many microorganisms will grow slowly at these temperatures and eventually alter the taste and appearance of foods, pathogenic bacteria will not. Exceptions are the clostridia causing type E botulism, *Listeria monocytogenes*, and *Yersinia enterocolitica*. Freezing does not kill bacteria outright in significant numbers, but frozen bacterial populations become dormant and decline slowly with time. Some parasites such as the trichinosis roundworm are killed by several days of freezing.

Radiation and Food Preservation

The use of **ionizing radiation** to preserve food is theoretically possible. Actual sterilzation by radiation probably causes too much alteration in the taste of most foods for commercial use. The use of radiation to treat spices, eliminate insects in some foods, and to kill trichinosis worms in pork, recently has been approved. **Microwaves** kill bacteria only by their heating effects on food.

Chemical Preservatives

Sorbic acid and **sodium benzoate** prevent mold growth in certain acidic foods such as cheese and soft drinks. Such foods are susceptible to mold-type spoilage. **Calcium propionate** is a fungistat used in breads. It prevents the growth of surface molds and the *Bacillus* species of bacteria that causes ropy bread. These organic acids do not work by any pH effect in inhibiting mold growth, but rather interfere with mold metabolism or the integrity of the cellular membrane. They are readily metabolized by the body.

Sodium nitrate and **sodium nitrite** are found in meat products such as ham, bacon, wieners, and sausage. The active ingredient is sodium nitrite; the sodium nitrate is a reservoir that some bacteria use as a substitute for oxygen—producing nitrites in the process. The nitrite preserves the red color of meat by reacting with blood components, and prevents the germination and growth of any botulism endospores. There has been some concern that reaction of nitrites with amino acids forms carcinogenic **nitrosamines**, and that this result might be a reason for removing nitrates from foods. It has been concluded, however, that the total removal of nitrates would not be justified because their value balances the low risks involved; however, nitrate amounts in foods have generally been reduced.

Food-Borne Infections and Microbial Intoxications

Some food-borne diseases are caused by a pathogen carried in contaminated food to the host, where it grows and produces damaging toxins. Such food-borne infections are exemplified by gastroenteritis, typhoid fever, and dysentery. In microbial intoxications, the toxin is formed by microbial growth in the food and then ingested with the food. Examples of such diseases are botulism and staphylococcal food poisoning.

Dairy products, often consumed without cooking, are particularly likely to transmit diseases. Standards are therefore strict. **Pasteurized Grade A milk** is required to have a standard plate count of less than 20,000 bacteria per milliliter and not more than 10 coliforms per milliliter. **Cultured** products such as buttermilk and cultured sour cream are required to contain no more than 10 coliforms per milliliter. **Grade A dry milk** products must have a standard plate count of less than 30,000 bacteria per gram or 10 coliforms per gram. **Certified raw milk** standards specify no more than 10,000 bacteria per milliliter or 10 coliforms per milliliter.

Roles of Microorganisms in Food Production

Cheese

Cheeses come in many types, all requiring the formation of a curd, which can then be separated from the liquid fraction, or **whey**. Except for unripened cheeses such as ricotta and cottage, the curd undergoes a microbial ripening process. The curd is milk protein, **casein**, and is usually formed by action of the enzyme, **rennin**, aided by acid conditions that are provided by inoculation with lactic acid-producing bacteria. These lactic bacteria provide the flavor and aroma.

Cheeses generally are classified by hardness. Romano and parmesan cheeses are very hard cheeses; cheddar and Swiss are hard; limburger, blue, or roquefort cheeses are semisoft; and camembert is a soft cheese. Hard cheddar and Swiss cheeses are ripened by lactic bacteria growing anaerobically in the interior. The longer the incubation time, the more acidity and the sharper the taste. (The holes in Swiss are from carbon dioxide produced by a *Propionibacterium* species of bacteria.) Semisoft cheeses are ripened by bacteria and other organisms growing on the surface. Blue and roquefort cheeses are ripened by *Penicillium* molds inoculated into the cheese. The open texture allows adequate oxygen to reach the molds. Camembert cheese is made in small packets so the ripening enzymes of surface molds will diffuse into the cheese.

Other Dairy Products

Butter is made by churning cream until the fat phase forms globules of butter separated from the liquid buttermilk fraction. Lactic acid bacteria are allowed to grow in the cream and provide the flavor from the **diacetyls** that give the typical butter flavor and aroma. **Buttermilk** usually is made by inoculating skim milk with bacteria that form lactic acid and diacetyls.

Yogurt is made from low-fat milk by first evaporating much of the water in a vacuum pan. This is followed by inoculation with a species of lactic acid-producing *Streptococcus* that grows at elevated temperatures. Stabilizers are added to aid in the formation of the thick texture. **Kefir** and **kumiss** are beverages of Eastern Europe. In these, acid-producing bacteria are supplemented with a lactose-fermenting yeast to give the drinks a small alcoholic content.

Nondairy Fermentations

Yeasts are used in baking; the sugars in bread dough are fermented to ethanol and carbon dioxide. The carbon dioxide makes the bubbles of leavened bread, and the ethanol evaporates in baking. Fermentation also is used in making sauerkraut, pickles, and olives. In Asia, soy sauce is a popular fermented food.

Beer and **ale** are products of fermentation of grain starches by yeast. The starch is first converted to glucose and maltose by a process called **malting**. Barley is allowed to sprout, then dried and ground. This results in malt, which contains amylases to degrade starch. **Distilled spirits**, such as whiskey, vodka, and rum, are made by fermentation of carbohydrates from grains, potatoes and molasses, respectively. The alcohol is then distilled off.

Wines typically are made from the yeast fermentation of grapes, which usually need no additional sugar—although more can be added to make more alcohol. Malic acid production makes some wines too acid, and bacteria are used to convert the malic acid to the less acidic lactic acid (**malolactic fermentation**). Wine can be spoiled by aerobic bacteria that convert alcohol into acetic acid. This process is used to make **vinegar**. Ethanol is made by the anaerobic fermentation of carbohydrates by yeasts. The ethanol is then aerobically oxidized to acetic acid by *Acetobacter* and *Gluconobacter* bacteria.

Microorganisms as a Food Source

In terms of world nutrition, protein is in particularly short supply. Microorganisms may become more important to the food supply because they can convert simple substrates into **single-cell protein (SCP)** very rapidly. Cellulose, methanol, petroleum hydrocarbons, and many industrial wastes might serve as substrates. Because photosynthesis is an inexhaustible energy supply, algae are attractive sources of SCP but are somewhat unpalatable. Bacteria are the most likely organisms from which to make SCP; they have a faster reproductive rate than fungi and can use more substrates than yeasts. High nucleic acid contents may be a problem in SCP because they may precipitate ailments such as gout.

Methanol is a good substrate for protein production because it is soluble in water, free of impurities, and readily separated from the end product. The bacterium *Methylophilus methylotrophus* is being grown now on methanol to produce protein animal feeds.

Industrial Microbiology

Industrial microbiology—the use of microbes to make industrial products—has centered on the lactic-acid and ethanol fermentations. In the future, as petroleum becomes more scarce, we may see a return to many microbial fermentations. In the coming years there will be a revolution in the application of genetically engineered microorganisms in industry, **biotechnology**.

Industrial Products

Amino Acids

Certain **amino acids**, such as **lysine**, are especially important because animals cannot synthesize them. They are present only in low levels in vegetable proteins and are therefore a valuable product. One bacterium that is used to produce lysine is *Corynebacterium glutamicum*, which normally produces both threonine and lysine. A mutant form of the bacterium that lacks the enzymes necessary to make threonine is used to produce lysine in large amounts. Enough threonine is added to the medium to allow this mutant organism to grow, but not enough is added to trigger feedback inhibition, which stops the production of lysine. The mutant produces lysine continuously, because lysine alone does not cause feedback inhibition. When this same bacterial species is provided with only a minimal amount of the vitamin biotin, it can be induced to excrete **glutamic acid**, which is used to make the flavor enhancer, monosodium glutamate.

Citric Acid

Citric acid is produced industrially by the mold, *Aspergillus niger* from molasses. This happens when the mold is provided with only a limited supply of iron and manganese.

Enzymes

Enzymes are valuable industrial products; **amylases** for syrups and paper sizing, **glucose isomerase** to convert glucose into fructose, **proteases** for baking, **proteolytic** enzymes for meat tenderizers, and additives to detergents.

Vitamins

Microbes produce several commercial vitamins. **Vitamin B$_{12}$** and **riboflavin** are two examples.

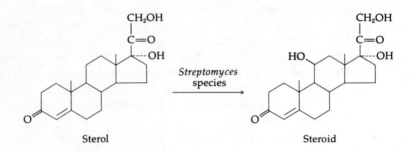

Figure 27-1 Production of steroids. Conversion of a precursor compund such as a sterol into a steroid by *Streptomyces*. These changes are extremely difficult to obtain by synthetic chemistry techniques, so microorganisms are often essential for economic production of pharmaceutical steroids.

Pharmaceuticals

Antibiotics are an important industrial product. Many are still produced by microbial fermentations. Work continues on selection of more productive mutants and manipulations to increase efficiency. **Vaccines** also are an important industrial product. **Steroids**, such as cortisone, and estrogens and progesterone used in birth control pills are products of microbial conversions (see Figure 27-1).

Uranium and Copper

The recovery of otherwise unprofitable grades of **uranium** and **copper ores** is made possible by the activity of the bacteria, *Thiobacillus ferrooxidans*. Water is sprinkled over a pile of the ore, and the bacteria change insoluble copper compounds to an oxidized, soluble form that moves out of the ore piles and can be reclaimed.

Microorganisms

Baker's yeast is a good example of microbes themselves as an industrial product. Others are the symbiotic nitrogen-fixing bacteria, *Rhizobium* and *Bradyrhizobium*, which are used to inoculate leguminous plants. The insect pathogen *Bacillus thuringiensis* is commonly sold in preparations applied to garden plants and trees.

Alternative Energy Sources Using Microorganisms

Conversion of biomass (various organic wastes) into alternative fuel sources is called **bioconversion**. **Methane** is an important source of such a process. It is produced even today in commercial amounts from large landfills and cattle-feeding lots. **Ethanol** represents an important additive to motor fuels in many parts of the world.

Industrial Applications of Genetic Engineering

The basic techniques of genetic engineering making use of the recombination of DNA molecules was discussed in Chapter 8. DNA is enzymatically cut up, and selected fragments are incorporated into plasmids, for example, which serve as a vector or carrier. These plasmids can be inserted into microorganisms, and gene amplification can increase the plasmids many times in a single cell. Short sequences of DNA used in genetic engineering can be produced by machine synthesis.

   Protoplast fusion is a technique that can be used to recombine DNA of species that normally do not exchange genetic information. This also is useful in the transfer of genetic information between plant species. Table 27-1 lists a representative group of genetically engineered products now being manufactured or under final development.

Table 27-1 Some Products of Genetic Engineering

Product	Comments
Pharmaceutical Products	
Tissue plasminogen activator(t-PA)	Dissolves the fibrin of blood clots; therapy for heart attacks
Erythropoietin (EPO)	Treatment of anemia
Insulin	Produced by *E. coli*; therapy for diabetics; this human insulin is better tolrated than insulin extracted from animals
Interleukin-2 (IL-2)	Produced by *E. coli*; possible treatment for cancer; stimulates the immune system
Gamma-interferon	Produced by *E. coli*, *Saccharomyces cerevisiae*; possible cancer, virus-disease therapy
Tumor necrosis factor (TNF)	Causes disintegration of tumor cells
Human growth hormone (HGH)	Produced by *E. coli*; corrects growth deficiencies in children
Epidermal growth factor (EGF)	Heals wounds, burns, ulcers
Monoclonal antibodies	Fusion of cancer cell and antibody-producing body cell; possible therapy for cancer, transplant rejection; diagnostic tests
Hepatitis B vaccine	Plasmid with hepatitis gene inserted into vaccinia, which is readily cultivated, whereas the hepatitis virus is not
Agricultural Products	
Ice-minus bacterium, *Pseudomonas syringae*	Lacks normal protein product that initiates undesirable ice formation on plants
Pseudomonas fluorescens bacterium	Has toxin-producing gene from insect pathogen *Bacillus thuringiensis*; toxin kills root-eating insects that ingest bacteria
Rhizobium meliloti bacterium	Modified for enhanced nitrogen fixation
Animal Husbandry Products	
Porcine growth hormone (PGH)	Improves weight gain in swine
Bovine growth hormone (BGH)	Improves weight gain in cattle
Food Production Products	
Rennin	Produced by *Aspergillus niger*; formation of milk curds for dairy products
Recreation Products	
Snomax®	*Pseudomonas syringae* makes a protein that causes water to freeze at higher-than-normal temperatures; used for making snow at ski resorts

Test

In the matching section, there is only one answer to each question; however, the lettered options (a, b, c, etc.) may be used more than once or, perhaps, not at all.

I. Match

1. Involved in making a less acidic wine.

2. Possible carcinogen in meat products.

3. Enzyme to form milk curd.

4. Responsible for flavors and aroma of dairy products.

5. Resistant bodies found on a number of molds.

6. An essential amino acid.

7. A product made by *Aspergillus niger* from molasses

a. Sclerotia

b. Malolactic fermentation

c. Nitrosamines

d. Rennin

e. Diacetyls

f. Citric acid

g. Lysine

II. Match (canned food spoilage types):

1. Caused by a thermophilic *Bacillus* species.

2. Swollen can; contents would have low pH and a sour odor.

3. Botulism would be a possible consideration in this type.

4. Probable cause would be can leakage or underprocessing.

a. Flat sour

b. Mesophilic

c. Thermophilic anaerobic

III. Match (food chemicals):

1. A fungistat used in breads to inhibit molds and ropy bread.

2. The active ingredient in many meats that preserves the red color and inhibits germination and growth of botulism organisms.

3. Used to prevent mold growth in certain acidic foods.

a. Sodium benzoate

b. Sodium nitrite

c. Sodium nitrate

d. Calcium propionate

IV. Match

1. Typhoid fever.

2. Staphylococcal food poisoning.

3. Botulism.

a. Food-borne intoxication

b. Food-borne infection

V. Match (milk microbiological standards per milliliter or gram):

1. Grade A pasteurized.

2. Dry milk.

3. Certified raw milk.

4. Cultured sour cream.

a. 20,000 bacteria/10 coliforms

b. 10,000 bacteria/10 coliforms

c. 10 coliforms

d. 30,000 bacteria/10 coliforms.

VI. Fill-in-the-Blanks

1. Cottage cheese is classified as a(n) _____ cheese.

2. Cheddar cheese is classified by hardness as a(n) _____ cheese.

3. Blue and roquefort cheeses are ripened by _____ molds.

4. The first step in making _____ is to evaporate much of the water from low-fat milk in a vacuum pan.

5. To make vinegar, ethanol is first made by an anaerobic fermentation of carbohydrates by _____. The ethanol is then aerobically oxidized to _____ acid by _____ bacteria.

6. One potential problem in the use of single-cell protein is high _____ content, precipitating ailments such as gout.

7. Glutamic acid is a microbial product used to make the flavor enhancer _____.

8. _____·fusion is a technique used to recombine DNA of species that normally do not exchange genetic information.

9. Steroids can be made by microbial alteration of available _____.

10. Canned food undergoes the minimum processing needed to destroy any *Clostridium botulinum* endospores. This is called _____.

11. In this type of processing (question 10), the population of any botulism bacteria would be decreased by _____ logarithmic cycles (give number).

12. Pathogens generally do not grow at refrigerator temperatures; one exception is the organism _____.

13. One parasite that is destroyed by several days freezing is the roundworm that causes _____.

14. Mild heating, short of sterilization, to destroy unwanted spoilage and pathogenic microorganisms, is called_____.

15. The liquid fraction remaining in milk after curd formation is called _____.

16. The holes in Swiss cheese are from carbon dioxide produced by a _____ species of bacteria.

17. A dairy food made, in part, by inoculation with a species of lactic acid-producing *Streptococcus* that grows at an elevated temperature is _____.

18. Packaging materials used in aseptic packaging are sterilized in a bath of hot _____.

19. Microwaves kill bacteria only by their _____ on food.

20. Pasteurization was first developed by _____.

Key

Matching

I.   1.b 2.c 3.d 4.e 5.a 6.g 7.f
II. 1.a 2.c 3.b 4.b
III. 1.d 2.b 3.a
IV. 1.b 2.a 3.a
V.   1.a 2.d 3.b 4.c

Fill-in-the-Blanks

1. unripened   2. hard   3. *Penicillium*   4. yogurt
5. yeasts; acetic; *Acetobacter* or *Gluconobacter* bacteria   6. nucleic acids
7. monosodium glutamate   8. protoplast   9. sterols   10. commercial sterilization   11. 12
12. *clostridium botulinum*, Type E, *Yersinia enterocoliticum*, or *Listeria monocytogenes*
13. trichinosis   14. pasteurization   15. whey   16. *Propionibacterium*   17. yogurt
18. hydrogen peroxide   19. heating effects   20. Louis Pasteur